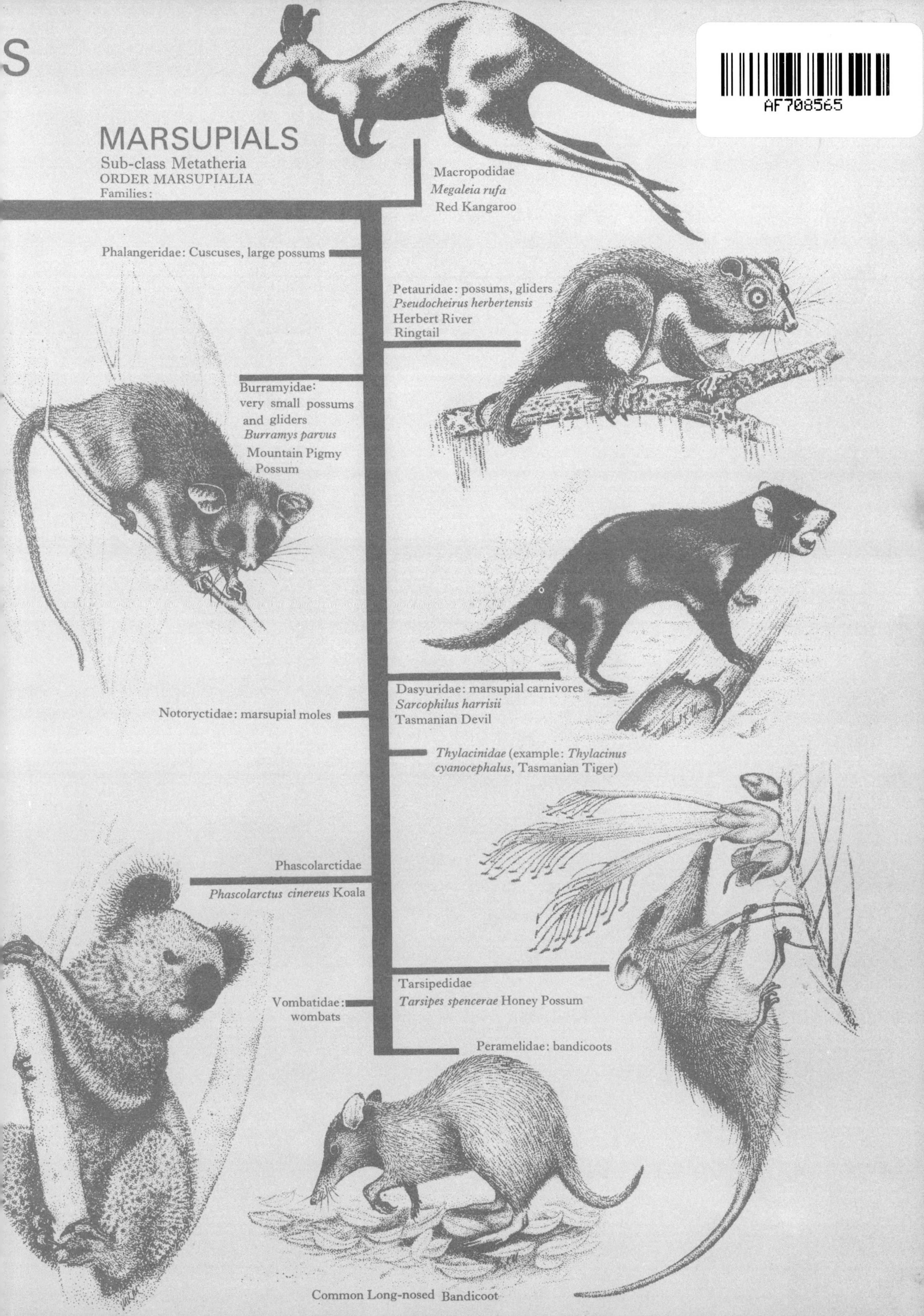
S
MARSUPIALS
Sub-class Metatheria
ORDER MARSUPIALIA
Families:
Macropodidae
Megaleia rufa
Red Kangaroo
Phalangeridae: Cuscuses, large possums
Petauridae: possums, gliders
Pseudocheirus herbertensis
Herbert River
Ringtail
Burramyidae:
very small possums
and gliders
Burramys parvus
Mountain Pigmy
Possum
Dasyuridae: marsupial carnivores
Sarcophilus harrisii
Tasmanian Devil
Notoryctidae: marsupial moles
Thylacinidae (example: Thylacinus
cyanocephalus, Tasmanian Tiger)
Phascolarctidae
Phascolarctus cinereus Koala
Tarsipedidae
Tarsipes spencerae Honey Possum
Vombatidae:
wombats
Peramelidae: bandicoots
Common Long-nosed Bandicoot

Rec'd 17 October 1983, from Paul Gritis, Cicero, Ill #17.00

Larry C. Watkins

Michael Morcombe's

AUSTRALIAN MARSUPIALS AND OTHER NATIVE MAMMALS

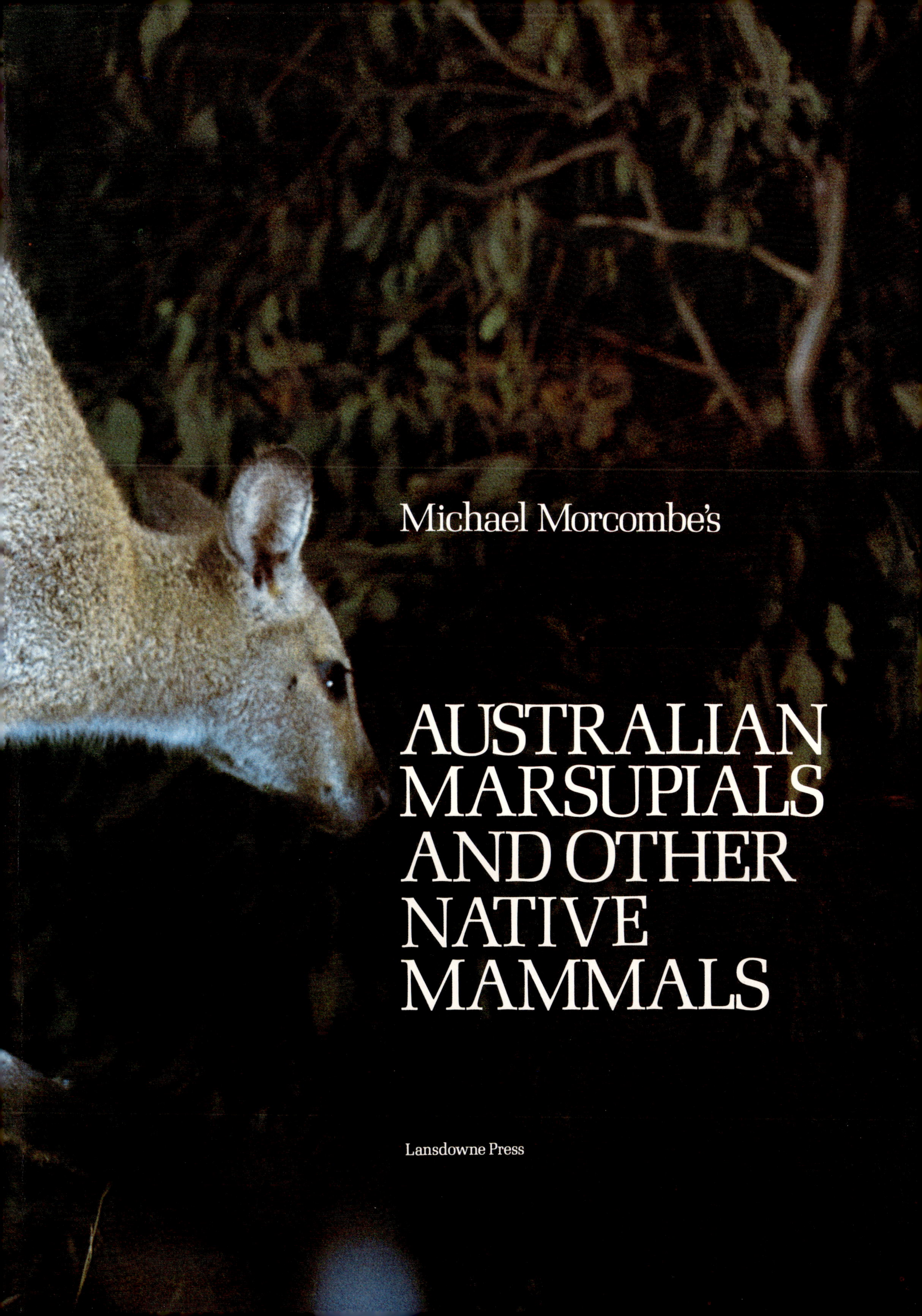

Michael Morcombe's

AUSTRALIAN MARSUPIALS AND OTHER NATIVE MAMMALS

Lansdowne Press

First published 1972
Lansdowne Press Pty Ltd
37 Little Bourke Street Melbourne 3000

Type set by Dudley E. King, Linotypers Melbourne 3000
Printed and bound in Hong Kong
SBN 7018 0275 8

During the preparation of this book, including some photography of mammals extending back over a period of five years, many people in various parts of Australia have played a part in one way or another, especially in the finding of various species which otherwise might not have been included: Frank Bailey, John Bannister, Alex Baynes, Miss Joan Bree, Rob Breeden, Dr M. Christian, Mr and Mrs J. P. Conquest, Ian Edgar, Mike Ellis, Bob Ferris, Doug Gilmour, Jim Hargreaves, Bernie Hyland, Ron Johnstone, Dr J. Kirsch, Malc. Lewis, Roly Paine, John Penrose, Greg Perry, Dr W. D. L. Ride, Ted Rotherham, Tom Spence, Kevin Sparkes, Bert Wells and Eric Worrell.

CONTENTS

Below
Desert Kangaroo-mouse or Hopping-mouse

INTRODUCTION

As a fauna region Australia is truly distinct from the rest of the world. It has been cut off by sea from the continents to the north for some fifty million years. Early primitive mammals which were marooned on this island continent found it a sanctuary free of competition from the more advanced types of mammals that were appearing on other continents.

Eventually marsupials vanished almost entirely from the rest of the globe, replaced by the proliferating placental mammals, but isolated Australia remained a marsupial-dominated sanctuary—a living museum of early forms of mammals.

Australia and Asia are almost linked by a chain of islands that extends westwards from New Guinea through Indonesia to Malaya. Those of the islands that lie in shallow seas on the Asian and Australian continental shelves have been linked at times when, in the great ice ages, the level of the oceans fell. Yet the line of demarcation between Australian and Asian faunas has remained surprisingly distinct. Known as Wallace's Line, after the first zoologist to comment on the clear-cut separation of Asian and Australian faunas at this point, the north-south line follows a deep ocean channel west of the Celebes and Lombok Islands. Borneo and Java are the first islands on the Asian side, inhabited by such placental mammals as deer, shrews, cats, squirrels and monkeys.

But eastwards of that dividing line the fauna of the islands becomes more dominantly Australian in character, with marsupials such as the cuscuses replacing equivalent placentals, the monkeys. Some species have crossed far beyond this point in either direction along the island chain, but nevertheless those deeper straits between the islands of the Indonesian archipelago have kept the Australian fauna remarkably distinct and intact over many millions of years.

Those placental mammals which did manage to reach and colonise Australia achieved this only comparatively recently, and were mostly small animals. Although they have, since arrival, evolved a great number of exclusively Australian species, their recent descent from mammals of other parts is obvious from their similarity. These late arrivals, some much later than others, include the native mice and rats, the bats and flying foxes, and the dingo which probably was brought down the island chain by wandering native peoples.

Within Australia the mammals—whether monotremes, marsupials or placentals—have encountered a great diversity of environmental conditions.

In spreading across the face of the continent to inhabit those vastly differing regions, the first marsupials, and later the placental mammals, became in time separated from others of their own kind, each isolated population gradually acquiring characteristics to suit the local habitat, and eventually becoming different enough to be distinct new species.

Over the millions of years the mammals have diversified to fill almost every available habitat from jungle canopy to desert dunes until, by the time of the arrival of European man, there were about 120 species of marsupials, 108 species of native placental mammals, and two species of monotremes in Australia.

Australia and New Guinea are the only places on earth where the three basic types of mammal exist side by side; the greatest differences between them lie in their way of reproduction. With marsupials the young are born at an age and size which seems incredibly premature, because there is no provision within the womb for nourishment of the foetus once it has grown beyond this size. Immediately after birth, the tiny young one struggles unaided through the fur to the pouch where it becomes firmly attached and remains for a long time.

The placental mammals, however, have succeeded in perfecting a device that enables the foetus to remain in the perfect environment of the womb for the full term. This structure is the placenta, which brings bloodstreams of mother and developing young into contact for exchange of the elements essential for respiration and growth.

The young of the largest marsupial, the Red Kangaroo, is only three-quarters of an inch long at birth, but some young placental mammals (such as foals and fawns) can run with their mother soon after birth. The superiority of the placental mammals lies in the shorter total time required, enabling them to multiply faster than most comparable marsupials.

The third major group of mammals are the monotremes, of which there are only three species including one in New Guinea. These animals are considered to be relics of ancient origin, and very primitive in their similarity to reptiles, especially in the egg-laying way of reproduction. They are, however, extremely specialised—the platypus to its amphibious way of life and the spiny anteaters in their ant diet and protective spines—and it is probably because of their success in this specialisation that they have never been replaced by marsupials.

Right

The female Red, or Plains Kangaroo, and her joey

THE MARSUPIAL CARNIVORE:

Native-cats, Marsupial-mice and the Numbat

The marsupial carnivores were probably the first Australian marsupials. From these unspecialised primitive mouse and rat-like insect-eaters have evolved not only the larger canivores, but also the whole multitude of herbivorous marsupials.

Although the vast majority of the marsupials in Australia today are quite small, and prey principally upon insects, there are some quite large predators: the Tiger Cat, the Tasmanian Devil and the Tasmanian Wolf or Tiger. In past ages there were some much bigger carnivores, including a 'Marsupial Lion', *Thylacoleo*.

Many of the small insectivorous marsupials such as the Dunnart are easily mistaken for mice or rats, but any similarity is quite superficial. The alert little marsupial-mice have sharp-pointed faces when compared with the blunt-nosed faces of the rodents, and they have white needle-pointed cat-like teeth instead of the yellowish chisel-shaped teeth of ordinary rats and mice. Being marsupials they carry their young in a pouch, which generally has not yet evolved to becoming a deep pocket but is merely a shallow depression which contains the teats where the young remain firmly attached for a long period after their birth.

The Numbat, or Marsupial Anteater, is a unique, highly specialised carnivore which feeds almost entirely upon termites. It has a long ribbon-like extensile tongue for probing the termites out of their narrow galleries, and a great many very small degenerate teeth.

Settlement and development have greatly thinned out our marsupial carnivores, and brought some species very close to extinction. The Tasmanian Tiger is extremely rare if not extinct, the Tiger Cat is rare in most areas except Tasmania, the Eastern Native-cat is rare on the mainland, the Dibbler or Freckled Marsupial-mouse may be nearly extinct, and the Numbat has long since gone from eastern Australia.

The Tiger Cat, *Dasyurus maculatus*, largest marsupial carnivore inhabiting the Australian mainland, grows to a length of almost four feet including the spotted tail. Equipped with long claws and jaws that open tremendously wide, it is a powerful and deadly hunter, preying upon birds and small mammals both in trees and on the ground.

Left

The Tiger Cat inhabits the dense eucalypt forests and rainforests of eastern coastal Australia and the mountain ranges; it is more common in Tasmania. Nocturnal and extremely wary, it carries its young in a pouch, a crescent-shaped flap of skin unlike the familiar deep pouch of the kangaroos.

Below

The Tasmanian Devil is probably the most ferocious looking, and sounding, of all Australian mammals. Although usually less than four feet in length, the Devil is extremely solidly and powerfully built. In the days of early settlement of Tasmania it acquired a reputation as a sheep killer. The Devil, *Sarcophilus harrisii*, has become more common in Tasmania in recent years; it became extinct on the mainland prior to European settlement.

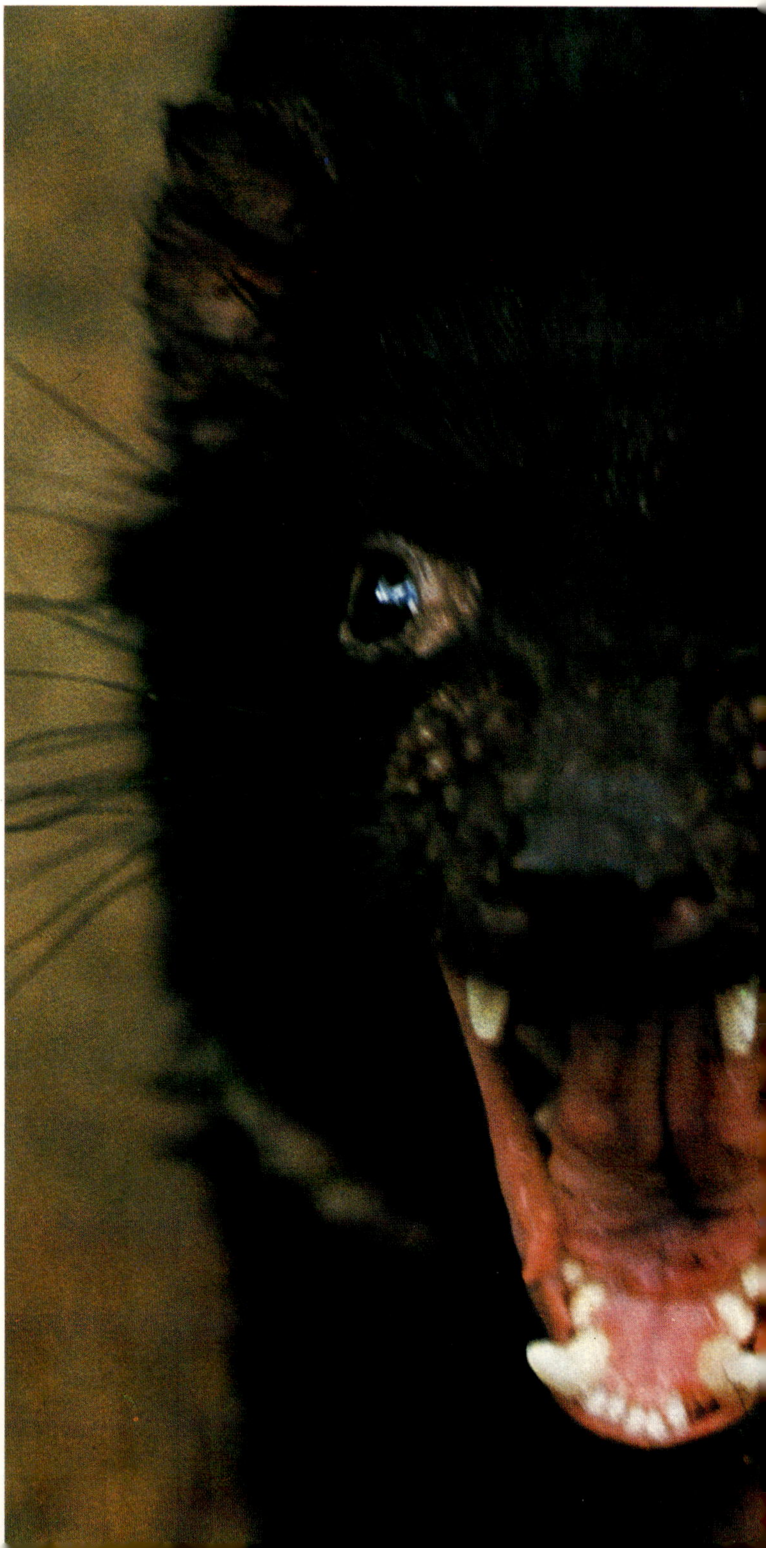

Below

The Chudich or Western Native-cat, *Dasyurus geoffroii*, is much smaller than the Tiger Cat; being only about thirty inches in length included the unspotted tail. With claws and teeth it rips apart a parrot which it has caught by surprise at night. Formerly found through most of inland and western parts of the continent, it now seems to exist only in the forests of south-western Australia. Two similar native-cat species inhabit the forests of eastern and tropical northern Australia; unlike the Tiger Cat, these native-cats have spotted tails.

After dark the Fat-tailed Dunnart emerges from a knot-hole in a small hollow log. At first it stands upright, looking cautiously in all directions, then sneaks away through the grass in search of grasshoppers, spiders and other small creatures. These it will attack with all the ferocity of a miniature native-cat, though it is only five inches long, tail included. This thickened tail is used for food storage and assists survival through drought or other hard times. The Dunnart, *Sminthopsis crassicaudata*, occurs in all States, except Tasmania, mostly inland in rather dry regions.

Left & Below

Unlike most marsupials, the Numbat or Marsupial Anteater is out and about during daylight hours. It may be glimpsed dashing across a bush track, tail arched high above its back, or observed scratching around a termite mound for the white ants upon which it feeds almost exclusively. Occasionally, if it hears a leaf rustle or a twig snap, the Numbat will stand upright on hind legs, ready at the first sign of danger to dash for the safety of one of the many hollow logs that litter the ground in its natural wandoo gum woodland home. Probably the most beautifully coloured of all Australian mammals, the Numbat, *Myrmecobius fasciatus*, once occurred as far east as New South Wales, but is long extinct in the eastern parts of its range, and now confined to south-western Australia except for a colony in the north-west of South Australia.

Below

As with all marsupials, Numbats are born incredibly small and, like the kangaroo's joey, they must climb through the fur to the pouch. The Numbat, like most of the marsupial carnivores, does not have a deep pocket-like pouch but merely a shallow depression in which the teats are grouped. The young Numbats cling with mouth and forepaws among the coarse fur, rarely moving while so very small and never at this stage releasing their grip. The female Numbat seems unaware of their presence, and certainly pays no attention to them. Later when they are much larger and able to release their rigid grip she will leave them in a nest burrow or hollow log while she is out feeding.

Below

In an alert, attentive pose, the female Numbat stands with one paw raised, a posture a Numbat can often be seen to adopt when, running through the bush, it suddenly sees or hears something strange and stops dead, as if frozen in mid-stride. Behind her rise the beautiful white wandoo gums which are characteristic of Numbat country. The tiny young, clinging among the fur beneath her body, can scarcely be seen.

15

Right
A Quenda, or Brown Bandicoot. This species belongs to the short-nosed group of bandicoots. Although its nose looks long, it is quite short compared with the slender snouts of the long-nosed bandicoots. The nocturnal Quenda, *Isoodon obesulus*, digs little conical pits in search of insects and other small invertebrates.

Below & Far Right
'Dibbler' was the native name for this speckled-furred marsupial-mouse of which a number of specimens were collected before 1884. During the next eighty-three years not one Dibbler was found. It was rediscovered in 1967 when the author trapped two Dibblers on the south coast of Western Australia. It was found that this insectivorous marsupial, which is about nine inches in length, climbs on to the flowers of the low coastal banksia trees in summer in search of nectar and insects. Recent attempts to locate more Dibblers have been unsuccessful, and it remains one of Australia's rarest mammals.

CUSCUS, KOALA AND WOMBATS

These interesting and well-known marsupials share with the possums certain features of feet and teeth that show them to be more closely related to the kangaroos than to the marsupial carnivores. Together with possums and kangaroos, they form the largest group of marsupials, the Diprotodonta.

The monkey-like cuscuses have come to Australia from New Guinea, where there are many species. Cuscuses, together with possums and gliders, belong to the family Phalangeridae, identified by the prehensile tail and hand-like hind foot.

The Koala, although now highly specialised for life in the treetops, has at some stage of evolution lost the tail which is so useful to arboreal mammals. It is thought that this may indicate a period of terrestrial life somewhere in its ancestry. The Koala appears to be more closely related to the wombats than to other marsupials. For example, it differs from the possums in having cheek pouches which are associated with the chewing of the bulky leaf food; in wombats there are similar though rudimentary cheek pouches. Both Koala and wombats have pouches which open backwards, but possum pouches open forward.

The destruction of forests, epidemics and the killing of thousands for their fur reduced Koala numbers to a dangerously low level.

Australia has four species of wombats, in two genera, the forests wombats, *Vombatus*, and the plains wombats, *Lasiorhinus*. The former genus, to which the Common and Tasmanian Wombats belong, is probably safe, inhabiting forests and rocky ranges over a wide area, but the various plains wombats (the species of hairy-nosed wombats) are mostly confined to restricted localities where they are very vulnerable to interference and therefore possible extinction.

The Spotted Cuscus inhabits the rainforests of tropical Queensland and New Guinea. It is quite a large animal, almost the size of a koala, nearly four feet in total length including the very distinctive tail. Covered with rasp-like scales, naked of fur on all sides for its terminal half and on its underside for the remainder of its length, the tail is fully prehensile, being able to encircle and grip branches. It is usually carried tightly rolled like the tail of the common ring-tailed possums. This tail is a characteristic of the cuscus genus, which is widespread in the tropical northern islands. The New Guinea Spotted Cuscus has a slightly longer tail and somewhat different colour compared with those populating Cape York.

Below
A female Spotted Cuscus, *Phalanger maculatus*, emerging from a hollow tree trunk where she sleeps during the day, carries a young one in her pouch. The females of both the Australian and New Guinea forms of the species are unspotted.

Left
With diminutive ears hidden in thick woolly fur and big night-seeing eyes, the slow-moving Cuscus has an appearance quite unlike other Australian arboreal mammals.

Right
The colour of the Spotted Cuscus varies greatly. The general colour of the males may be spotted or plain; the Australian sub-species is generally darker and greyer, while the New Guinea Spotted Cuscuses often have a yellowish or semi-albino coloration. Albino cuscuses seem fairly common. The size of the spots and blotches vary greatly, the Australian males having more extensive patches of the darker colour.

The highly specialised, almost exclusively arboreal Koala is one of Australia's best-known animals. It is specialised not only in its physical adaptations for a life in the tree tops but also in its diet. The Koala's hands have a vice-like grip between the first two and the other three fingers, and together with the very long arms and curved claws this enables it to climb smooth tree trunks and compensates for the lack of any useful tail. The Koala will eat only the leaves of a few species of smooth-barked eucalypts. Its young are carried in a backward-opening pouch (like that of the wombats) for five or six months, and on the back for a further two or three months. The Koala, *Phascolarctos cinereus*, inhabits eastern Australia from southern Queensland to Victoria and South Australia.

Right

The Tasmanian Wombats (and the almost identical Islands Wombat of the Bass Strait islands) belong to the forest or naked-nosed genus, and are the smallest of the wombats. They are now thought to be insular forms of the Common Wombat rather than distinct species.

Below

The Hairy-nosed Wombats of the Nullarbor Plains live in large colonies where many burrows form a complex of tunnels usually with the entrances meeting as a large crater. Although these wombats are most active at dusk and dawn they often emerge from their burrows in mild calm weather to lie in the sun.

Left

The Common Wombat, *Vombatus ursinus*, inhabits eastern Australia from south-eastern Queensland through eastern New South Wales and southern Victoria to eastern South Australia, in forested and usually rocky country. This and the Tasmanian and Islands Wombats can be separated from the three species of hairy-nosed wombats by their very coarse fur, short ears and bare snout.

Below

Lumbering from its burrow, a Hairy-nosed Wombat of the Blanchetown area of South Australia shows its distinctive snout. This and the wombat of the Nullarbor are slightly different geographical forms of the one species, *Lasiorhinus latifrons*, and easily distinguished from the Common, Tasmanian and Islands Wombats not only by their distinctive muzzle but also by their fine silky fur and long ears.

POSSUMS, GLIDERS, AND THE NOOLBENGER

These Australian mammals all live and feed in the treetops. Being closely related they have each inherited from some common ancestor certain adaptations for the arboreal way of life.

Each has a long tail which is either prehensile (and therefore able to curl around branches and grip more tightly than a hand); or bushy, fluffy or feather-like, which is used as a balance and rudder during long leaps and gliding flights.

But the most distinctive feature of the family Phalangeridae to which possums, gliders and cuscuses belong, is the hand-like shape of the hind feet. The clawless big toe, like a human thumb, is opposable to the rest of the foot, giving a firm grasp on the branches. There are strong claws on all other toes of hands and feet. The second and third toes, like those of kangaroos, are combined or syndactylous.

The possums may be considered in three main groups: the brushtails, the ringtails and the pigmy possums. Three species not fitting any of these groups are the Striped Possum, the Leadbeater's Possum and the Noolbenger or Honey Possum.

Australia's remarkable flying phalangers or gliders have evolved on three separate occasions, from three different groups of non-gliding possums. From the ringtails have been developed the Greater and Yellow-bellied Gliders, from the Leadbeater's Possum have come Sugar and Squirrel Gliders, and from the pigmy possums, the tiny Feathertail Glider.

The Honey Possum is a furred counterpart of the avian honeyeaters, whose nectar-eating it duplicates at night among the wildflowers of the south coast of Western Australia. Only in this region, with such a wealth and diversity of wildflowers, has such a marsupial evolved. With beak-like snout and fine, darting brush-tipped tongue it has an advantage over any other mammal that may try to reach the nectar provided by the flowers.

Right
The Bobuck, Mountain Brushtail or Short-eared Brushtail, *Trichosurus caninus*, has ears that are short and rounded compared with those of other brushtail species, and the fur has a glossy sheen.

Below
The common grey Brush-tailed Possum, *Trichosurus vulpeca*, of south-eastern Australia, has long pointed ears; its colour varies from grey to black. This is one of the commonest of marsupials, and one which has adapted exceptionally well to the changed environment, frequently living in the roofs of suburban homes

Below

A Brush-tailed Possum of the Mareeba district of north Queensland, it is so widely distributed that many different forms of it occur particularly towards the limits of its geographical range. Some seven different sub-species are recognised, as well as numerous slightly different colour phases of the common grey animal. Some of the sub-species are quite distinctive, such as the Coppery Brushtail of the Atherton Tableland, a large chestnut-brown possum. Other populations of Brush-tail isolated in Tasmania, south-western Australia and on Cape York are also slightly different to those of the south-east.

Right

The Bobuck or Short-eared Brushtail must be one of the most attractive of the large possums, with a silvery sheen to his luxuriant fur, a face that carries an alert, almost surprised expression, and big bulging brown eyes like those of a ringtail possum. His ears are about half the length of the ears of other brushtails. but he is equal in size to the largest forms of the Common Brushtailed Possum. The Bobuck is an inhabitant of the dense rainforests from south-eastern Queensland through New South Wales to eastern Victoria.

Left

Hanging by his back feet, an angry Northern Brushtail lashes out with sharp claws and attempts to use his teeth. This smaller, more slender brushtail, *Trichosurus arnhemensis*, occurs in the far north of Australia, from the Gulf of Carpentaria through Arnhem Land to the Kimberley.

Left

Hanging by his back feet, an angry Northern Brushtail lashes out with sharp claws and attempts to use his teeth. This smaller, more slender brushtail, *Trichosurus arnhemensis*, occurs in the far north of Australia, from the Gulf of Carpentaria through Arnhem Land to the Kimberley.

Leaping from one branch to another the Bobuck shows the agility of the brushtails, which are much faster moving through the tree tops than the cautious ringtail possums. His feet, outstretched, are clearly visible. The hind feet as well as the front are hand-like, and quite unlike the narrow elongated hind feet of kangaroos and bandicoots. With a first toe like a thumb, opposable, the hind foot is capable of grasping. This characteristic the brushtails share with the cuscuses, the Koala and the gliders. The tail is long and bushy, but naked on its undersurface, and with a small bare end like the tip of a finger.

Below

One of the strangest of the ringtail possums is north Queensland's Green Ringtail, or Striped Ringtail, *Pseudocheirus archeri*, which is found only in the rainforests of the Atherton Tableland and nearby mountain ranges. Its fur is a peculiar pale greenish-brown colour unique among Australian mammals, and there are two stripes down its back.

Right

The north Queensland form of Common Ringtail, *Pseudocheirus peregrinus incanens*, is another of the approximately eleven sub-species of the ordinary grey-brown ringtail. Although various geographical forms of the Common Ringtail have in the past been classified as distinct species, recent studies of their blood proteins strengthen the view that they are isolated populations of Common Ringtail which have in time evolved some differences, but insufficient to establish them as separate species.

Left

The Mongan or Herbert River Ringtail, *Pseudocheirus herbertensis*, is a beautifully patterned, chocolate-brown possum, with pure white underparts and the great bulging sensitive eyes a rich red. The young ones lack the crisp contrast of brown and white of the adult. This species, discovered in the jungles around the head waters of the Herbert River, is restricted to the mountain rainforests of north-eastern Queensland.

Below

Inhabiting the forests of south-western Australia is the unusually dark-furred Western Ringtail, which has usually been classified as a distinct species, *Pseudocheirus occidentalis*, but may be just another isolated race of the common eastern species *Pseudocheirus peregrinus*. This almost black ringtail, like the Common Ringtail, builds a large globular nest in peppermint scrub thickets or other dense vegetation, but also at times lives in hollows of trees.

Below

The Common Ringtail is widespread in forested coastal areas of Australia. As would be expected with such a wide distribution from tropical rainforests to southern mountain forests, it varies greatly in colour and other features. This is the south-eastern sub-species, *Pseudocheirus peregrinus laniginosus*.

Right

The most widespread of the gliders is the Sugar Glider, *Petaurus breviceps*. This glider is nocturnal, and feeds on insects and nectar. The Sugar Glider has been known to sail through the air a distance of fifty yards. It occurs in coastal forests of northern and eastern Australia from the Kimberleys to Adelaide, and has been introduced into Tasmania.

Below

Almost identical in appearance to the Sugar Glider, the Squirrel Glider, *Petaurus norfolcensis*, is considerably larger, having a total length of twenty inches. Its colour differs slightly, the undersurfaces being white to creamy-white instead of very pale grey to medium grey, and the dark dorsal line is usually more distinct when specimens of Sugar and Squirrel Gliders of the one locality are compared. The Squirrel Glider inhabits eastern Australian forests and woodlands from about Cardwell, Queensland, to Melbourne.

Below

Clinging to a tropical Queensland Erythrina tree a Feathertail Glider, *Acrobates pygmaeus*, pauses for a moment in its raiding of the flowers for their nectar. This mouse-sized phalanger, known also as the Pigmy Glider, inhabits the eucalypt forests of eastern Australia from Cape York to Victoria and the south-east corner of South Australia. Because it is completely nocturnal the Feathertail, by far the smallest of the gliders, is not often seen. It is evolved from the pigmy possums, and only distantly related to the other gliders.

Right

Gliding swiftly downwards from some higher branch the Feathertail Glider at the last instant swoops upwards, checking its speed almost to stalling point before dropping on to the cluster of eucalypt flowers. Clearly visible is its distinctive tail, which is fringed along each side with longer hairs like the vanes of a feather, and serves in flight as a balancer and rudder. The remarkable flight membrane is an extension of the side body skin, connected along the limbs out to wrists and ankles, so that it stretches tightly when the limbs are outstretched.

Left

The Greater Glider, about thirty to thirty-six inches in total length, inhabits coastal eastern Australia from Cairns to Melbourne. Shown here is the rather smaller northern sub-species, *Schoinobates volans minor*.

Below

Gliding down to a landing at the base of a tree a Greater Glider shows her outstretched parachute-like membrane; her pouch opening is clearly visible. At the end of the fast glide, which from a tall tree may cover a distance of up to eighty yards, the glider swoops upwards, body almost vertical, reducing speed and contacting the tree trunk in an upright position.

Left

The little Pigmy Possums have huge bulging black eyes and sensitive ears to guide them in their nocturnal hunting for insects and wildflowers.

Below

A mouse-sized South-western Pigmy possum, *Cercartetus concinnus*, washes its sticky fingers fastidiously after feeding among eucalypt flowers.

Over

The unique Honey Possum, *Tarsipes spencerae*, has evolved a snout like the beak of a honey-eating bird. It is so specialised for its nectar a diet that its teeth are mere vestiges. When the Honey Possum hangs from its feet and prehensile tail it shows the stripes down its back which make it easy to identify. Its lips and long snout form a tube, through which it can draw up nectar with a brush-tipped tongue. The Honey Possum is sometimes known by its native name, Noolbenger.

KANGAROOS AND WALLAROOS

The great kangaroos are not only the best known of the Australian mammals, but also the largest. Although their general appearance is familiar to everyone, identification of the many species and varied geographical races can be difficult. Fortunately there are usually certain characteristics which remain constant for each species even though the colour may vary enormously from one region to another.

The great kangaroos include the Red of the inland plains, the greys of southern forests and woodlands, and the wallaroos or euros which prefer rocky hills of both coastal and inland areas. Compared with the Red Kangaroo and the two species of grey, wallaroos look shorter and more heavily built, an impression resulting partly from their long shaggy fur. Closely related to the wallaroos is the Antilopine Kangaroo of the northern tropical grasslands.

Kangaroos are famed for their pouch way of carrying their young, but in this respect they are far from unique. Nor are they alone in their jumping, this way of movement being duplicated by the hopping mice. But these characteristics in mammals of such large size, together with widespread occurrence as the continent's dominant grazing animal, have attracted a great amount of popular attention to the kangaroos.

The kangaroos, together with the smaller wallabies, are known as the Macropodidae, a reference to the very large elongated hind feet. Their long and heavy tail serves as a balance when they are travelling at great speed. It also serves as an extra leg when standing or moving very slowly.

In common with other herbivorous mammals, the kangaroos have teeth modified for grazing. The little front teeth of the lower jaw have gone, leaving only two very large scissor-like protruding incisors which, in grazing, bear up against a leathery pad between the semicircle of the three pairs of upper incisors, an arrangement known as diprotodonty. Kangaroos, possums, the Koala and wombats all have a similar arrangement of teeth, and together form a large marsupial group known as the Diprotodonta.

Below
The Western Grey Kangaroo of the forested south-western corner of the continent has brownish-grey fur without any silvery sheen, and dark brown on the face and around the base of the ears. Formerly known as *Macropus ocydromus*, it is now considered to be a form of the South Australian and western Victorian species *Macropus fulginosus*, which it very closely resembles.

Right
This female Great Grey, *Macropus giganteus*, pauses in scrubby woodland to feed with her joey.

One of the largest of the kangaroos, the Great Grey or Forester, *Macropus giganteus*, may stand over six feet tall and weigh more than 180 pounds. The female is considerably smaller. Among the distinguishing features are the hairy snout, the absence of any black or white face marks, the fur short and woolly, grey to grey-brown with a silvery sheen, becoming almost white on the undersurface, and the grey-brown tail black at the tip. The Great Grey inhabits forests and open woodlands from northern Queensland to Victoria and Tasmania.

Right & Below

The joey first starts to leave the pouch at six months of age, and finally vacates the pouch about seven weeks later. Although no longer using the pouch, the joey continues to suckle from the outside until about one year old. Female Reds may be brownish, or grey when they are known as Blue Fliers or Blue Does.

Far Right

A huge male Red Kangaroo, stretched out in the shade of the scrub to escape the mid-day heat, shows the unvarying facial markings which make possible the positive identification of the species regardless of colour. Although known as 'the Red', males of this species, *Megaleia rufa*, may be entirely grey, or entirely reddish-brown, or reddish with a grey head. But whatever the colour, the conspicuous white and black muzzle markings are consistent. The end of the tail is a paler colour than the body.

Left
The Euro or Roan Wallaroo, *Macropus robustus erubescens*, is an intermediate sub-species occurring in South Australia, central Australia and Queensland, midway between the very dark blue-grey Wallaroo, and the Red Wallaroos.

Below
The Antilopine Kangaroo, *Macropus antilopinus*, is closely related to the euros, but has shorter, thinner fur. It bears a superficial resemblance to the Red Kangaroo (and grows almost as big) but lacks the Red's facial markings. The Antilopine inhabits the savannah-woodlands of tropical northern Australia.

Right
Immediately recognisable by its heavy build and long shaggy fur as one of the wallaroos, this powerful big male is further identified by his rich red colour as a Red Wallaroo or Western Euro, *Macropus robustus cervinus*, which inhabit the ranges of the north-west and interior of Western Australia.

5

Top Left
The Kangaroo Island Kangaroo, *Macropus fuliginosus fuliginosus*, living on a predator-free island, has become heavier, darker, and very slow-moving compared with closely-related mainland kangaroos.

Centre Left & Right
Red Wallaroo, or Western Euro, female.

Forester or Great Grey, *Macropus giganteus*, female.

Lower Left & Right
The Black-faced, Sooty or Mallee Kangaroo, *Macropus fuliginosus melanops*, of South Australia, western Victoria and south-western New South Wales.

A female Red Kangaroo, entirely grey except for white underparts and the facial markings typical of the species.

Below
A grey-faced male Red Kangaroo, *Megaleia rufa*.

WALLABIES, PADEMELON, BETTONG AND POTOROO

The kangaroo family contains a great many species ranging in size from seven-foot tall giants to little rat-kangaroos. The largest wallabies have a total length, nose to tail-tip, of about seventy-two inches (for the Red-necked Wallaby), but other than size and the marking of individual species there is no obvious difference between large wallabies and the kangaroos.

The wallabies are a very diverse group, having expanded into a bewildering array of species and sub-species that occupy almost every available habitat. Their wide range of sizes probably results from their environment. They inhabit more densely vegetated country than the kangaroos and, the smaller the wallabies, the thicker the undergrowth they choose for their home.

Among the most interesting and colourful of this group of marsupials are the rock-wallabies. There are two genera, about seven species and many slightly differing sub-species. Rock-wallabies—in scattered, isolated populations—inhabit rough ranges and rocky outcrops throughout Australia, and occur on a number of small rocky islands around the coast; they are not present in Tasmania.

'Pademelon' is the name applied to certain very small wallabies which favour thickets and undergrowth. Apart from small size, they are distinguished by their relatively short tails which look thicker, not as gracefully tapered as the tails of the wallabies, and there is a minor difference in dentition. Being small they are very vulnerable, and have become rare or extinct in many of their former haunts. One species still common is Tasmania's Red-bellied Pademelon.

The rat-kangaroo group contains the smallest members of the kangaroo family. Apart from the rather rat-like appearance, these marsupials are separated by their short rounded ears, and a more primitive dentition. In this group are included the bettongs or brush-tailed rat-kangaroos, which are still partly insectivorous, and the little bandicoot-like potoroos which have shorter back legs and use their front legs in a galloping run instead of hopping.

Right
The Agile Wallaby, also known as Sandy Wallaby and River Wallaby, *Macropus agilis*, is a common large wallaby of northern Australia, from North Queensland through the Northern Territory to the Kimberley. It is most often seen in river country and rich grassland plains where the grass grows fast and tall after the tropical wet season. During the day it prefers to shelter in gallery forest or other dense scrub. The Agile has very distinct cheek and hip stripes.

Below
The Whip-tail or Pretty-face Wallaby is a large slender animal with an exceptionally long tapering tail which is black at the very tip. There is a distinct white face stripe. This species, *Macropus parryi*, inhabits the woodlands and eucalypt forests of the coastal ranges of eastern Queensland and north-eastern New South Wales.

Left

The Swamp Wallaby, Black Wallaby or Black-tailed Wallaby prefers the moist gullies and thickets, or open forests if dense cover is available nearby. This species, *Wallabia bicolor*, occurs along the east coast from Queensland to Victoria and South Australia. Those from northern areas are more brownish above, tending to become rufous-orange on the underparts. There is an indistinct grey face stripe.

Below

In Tasmania the small wallabies and pademelons have survived much better than their mainland relatives, for the fox, which European man brought with him, is absent from that island. The Tasmanian or Red-bellied Pademelon, *Thylogale billardieri*, is still very abundant in Tasmania where it has a long dense shaggy coat in keeping with the colder climate.

Right
The Tammar or Dama Wallaby, *Macropus eugenii*, smallest of the wallabies, occurs in a number of isolated colonies in South Australia, south-western Australia and nearby islands including Kangaroo Island. Identification features include rufous shoulders, naked snout and faint dorsal stripe.

Below
Inhabiting the most rugged, rocky ranges, the Yellow-footed or Ring-tailed Rock-wallabies move with incredible agility and certainty. Their feet have soft, granular-surfaced pads to prevent slipping—the rock surfaces and ledges around the caves where they have lived for tens of thousands of years are polished smooth by countless generations of wallaby feet. This species, *Petrogale xanthopus*, inhabits the Flinders Ranges of South Australia and the ranges around Broken Hill; a sub-species occurs in south-western Queensland.

Left

One of the most attractive of rock-wallabies occurs on islands at the eastern end of the Great Australian Bight. The Pearson Islands Rock-wallaby, *Petrogale penicillata pearsoni*, has the usual rock-wallaby long bushy tail which serves as a balancer when it makes it spectacular leaps from one to another of the huge granite boulders that cover much of its island home.

Below

At full speed a Bennett's Wallaby leaps a fallen tree trunk, and a series of high-speed electronic flash exposures, each at 1/5,000th of a second, freeze it in a sequence of positions between take-off and landing. When travelling fast, a wallaby or kangaroo is airborne for about seventy percent of the time. Its hands and tail are held well clear of the ground, the heavy tail serving as a counterbalance to the head and shoulders as the great hind legs lash out with powerful spring action. The energy expended in fast travel prevents really sustained full-speed flight. But the erratic bounding action gives greater speed through forest undergrowth, dodging or leaping the bushes, logs, stumps and boulders; only on the smooth open plains is the macropod really at a disadvantage. Bennett's Wallaby, *Macropus rufogriseus rufogriseus*, occurs on King Island in Bass Strait. A similar looking sub-species occurs in Tasmania, and another (the Red-necked Wallaby) on the Australian mainland.

Left
One of the rarest of wallabies in Australia, the Parma or White-fronted Wallaby, *Macropus parma*, was thought to be extinct as none had been seen since 1932 but in 1965 it was rediscovered, not in Australia, but on Kawau Island off the coast of New Zealand, where it had been introduced long ago and forgotten. By remarkable coincidence, soon afterwards this species was found to be still surviving near Gosford, New South Wales.

Centre Left
The Agile Wallaby of the tropical north can be identified by its sandy-brown colour, short ears, snout partly hairy between the nostrils, and distinct white cheek and hip stripes.

Lower Left
Very similar to Bennett's Wallaby is the mainland sub-species, the Red-necked Wallaby, *Macropus rufogriseus banksiana*, which has a much more pronounced rufous tone to the fur of shoulders and rump. It inhabits the coastal forests of eastern and south-eastern Australia.

Lower Left

The Whiptail, or Pretty-face Wallaby, has pale grey fur with a very distinct white face stripe and dark grey around the base of the ears. The red glow in the eyes is a reflection of the flash light.

Below

Bennett's Wallaby can be distinguished from the Red-necked Wallaby by its more uniformly greyish colour, there being just a trace of the rufous tone on shoulders and rump.

Below

Dashing away through thick undergrowth a Brush-tailed Bettong or Woilie, *Bettongia penicillata*, shows its distinctive tail, which has longer hair towards the tip on the upper surface forming a black crest or brush. Once widely distributed across southern Australia, this rat-kangaroo is now extremely rare except in the wandoo forests of south-western Australia where it is quite common in places where suitable natural habitat remains.

The Southern Potoroo, *Potorous apicalis*, is one of the little rat-kangaroos which have become so rare in most parts of Australia following the destruction of much of the dense vegetation which is essential for protection and shelter. The depredations of the fox and the domestic cat gone wild have contributed to the loss or rarity of many of the smaller marsupials. This species occurs in coastal south-eastern Australia, where it is now rare, the Bass Strait islands and Tasmania. It inhabits rainforests, wet sclerophyll forests and the dense undergrowth around the margins of swamps.

63

THE MONOTREMES:
Platypus and Echidnas

The world's only living monotremes, the Platypus and the echidnas or spiny anteaters, are endemic to the Australian region. The Platypus is confined to eastern Australia, the Australian Echidna (*Tachyglossus*) occurs throughout the continent, and the Long-beaked Echidna (*Zaglossus*) inhabits New Guinea. They represent a prototype mammal which in its level of evolution stands little more than mid-way between the reptiles and the higher mammals.

Since reaching Australia probably well over a hundred million years ago, the ancestral monotremes, then not much above the early pre-mammal or mammal-like reptile stage, evolved very slowly, and up to the present time still retain many reptilian features. For reasons unknown, although representing an advance over the reptiles, the monotremes have never proliferated into the great diversity of species that later marsupial and placental mammals achieved. Possibly the early monotremes, with exception of Platypus and echidnas, were replaced by the more rapidly evolving and diversifying marsupials.

As there is comparatively little in the fossil records to explain these mysteries of the distant past, the three living species of monotremes must supply many of the answers. These 'living fossils' resemble reptiles in their skeletal structure, particularly of shoulder and hip regions, and of course in their reptilian egg-laying way of reproduction.

Reptiles are unable to control their body temperatures except by behavioural tactics such as basking in the sun, but mammals can maintain a steady warm body temperature by internal means. The monotremes have not quite achieved a control comparable to the advanced mammals, their temperature averaging somewhat lower at 88°F (31·1°C) and fluctuating between 72° and 96°F (22·2°C and 35·5°C).

But probably the greatest gain by the monotremes when compared with the reptiles is their much larger brain—as large as that of the more primitive of marsupial and placental mammals.

The Platypus emerges from its burrow in the early morning and late afternoon. In the water its webbed front feet expand into broad paddles. The hind feet, also webbed, trail behind, and together with the tail, give stability and directional control. The Platypus surfaces often to breathe, and to chew the prey captured underwater. When it dives the eyes and ears are closed. It seems that the prey is found and identified solely by touch, then snapped up and stored in cheek pouches.

Below

If it is disturbed, the echidna's immediate reaction is to burrow straight downwards, and once half buried, with the heavy claws of its immensely strong limbs hooked under roots or rocks, it is almost impossible to dislodge. But if the anteater does happen to be taken by surprise on a hard surface such as an area of unbroken rock, where it cannot get a secure grip, it curls into a tight ball, protecting its face and under-surface where there are no spines.

The echidna or spiny anteater, *Tachyglossus aculeatus*, although belonging to the most primitive group of mammals in the world, has remained one of the most successful of all Australian furred animals. It is extremely widely distributed throughout Australia and Tasmania, being common over a wide range of habitats from semi-desert to rainforest. The spiny anteater's long tube-like muzzle, like the 'beak' of the platypus, is an adaptation to a way of feeding. The tiny mouth opens just wide enough to allow the rapid probing action of the long tongue in picking up ants and termites. Confident of the protection given by its coat of spines, the anteater is out during daylight, and most active during early afternoon.

PLACENTAL MAMMALS

Native Rodents, The Dingo, Seals and Flying Foxes

So much attention is given to the marsupials and monotremes of Australia that the native placental mammals as a whole are rather a forgotten group. Although comparative newcomers, placental mammals reached this continent sufficiently long ago to have evolved a great many species that are unique to Australia.

Our placental mammals include animals as large as the Sea Lion and as small as the native-mice. Some, like the hopping mice, have adapted so well to the central deserts that they can live on dry seeds alone, drinking no water at all. Others, the water-rats, have joined the Platypus in the rivers and swamps, and now have webbed feet. Some of the Australian native-mice and native-rates have been so long isolated on this continent that they are referred to as the 'old endemics'. In this group are the hopping-mice, tree-rats, stick-nest rats, the rock-rats and the little native-mice of genus *Pseudomys*.

Identification of many of these is difficult and depends upon characteristics of skulls and teeth.

Bats are among the most numerous of Australian mammals, but not as noticeable as their numbers might suggest. They are the only mammals that fly. Bats can propel themselves through the air with wings that are modified forelimbs, long slender fingers like umbrella ribs supporting the delicate flight membranes.

Most of the Australian bats are small insectivorous species, but the Ghost Bat has a wingspan of two feet and captures birds and small mammals. The largest of the bats are the flying foxes, which feed on flowering trees and fruit.

Most of the Antarctic seals have been recorded around the Australian coastline. Four species breed on beaches of the mainland and the offshore islands: the Sea Lion, the Australian Fur Seal, the New Zealand Fur Seal, and the Elephant Seal.

The most recent of placental mammals to reach Australia was the Dingo, brought here by Aborigines such a comparatively short time ago that he can hardly be regarded as a true native animal. Like the fox, buffalo, rabbit, cat and others introduced by European man, the Dingo evolved on some other continent.

The Allied Rat or Southern Bush-rat of coastal Queensland, New South Wales and Victoria lives among dense low ground cover and debris of the forest floor. In its nocturnal wanderings if often climbs into the foliage of shrubs and trees in search of food, which consists of native vegetation.

Right
A Western Swamp Rat, *Rattus fuscipes*, feeding on a flowering *Banksia attenuata*.

Over

Australia has many native mice and rats related to ordinary mice and rats; yet most species are clean, shy, attractive little animals. Western Swamp-rats, *Rattus fuscipes*, are attracted to the nectar of banksia trees during hot dry summer months. This rodent lives in swampy areas, forests and sand-heath country in south-western Australia.

Below

One of the most attractive of the native rodents is the Paler Field-rat, or Tunney's Rat, a sandy-brown, long-furred, fluffy-looking short-tailed species. *Rattus tunneyi* occupies a very wide variety of habitats, including heathlands, grasslands, savannah-woodlands, coastal and river flats across the northern half of the continent, from north-eastern Queensland through northern and central Australia, to north-western Australia.

Right

Southern Bush-rat is now applied as a single collective name for three formerly distinct species, now combined together as the one species, *Rattus fuscipes*. The three species, one inhabiting forests of the eastern and south-eastern coasts, one the moister coastal parts of South Australia, and one the extreme south-west corner of the continent, had various common names: Allied Rat, Grey's Rat, and Western Swamp-rat. Recent studies suggest that these are but slightly different geographical races of the one species, so *Rattus assimilis* (pictured), *Rattus greyi* and *Rattus fuscipes* now become, together, *Rattus fuscipes*, that having been the first-named of the three.

Below

The majority of the Australian hopping-mice species inhabit northern and arid regions. They have evolved a two-footed leaping action like that of the kangaroos, and their general appearance is that of a tiny kangaroo, the size of a large ordinary mouse but with much longer hind legs and tail. Hopping-mice are able to live in desert regions because they avoid day-time heat, remaining in their burrows three feet or more down in the cool and usually slightly damp earth.

In the subterranean nest chamber the female Spinifex Hopping-mouse, *Notomys alexis*, washes her new-born young.

Far Below

A tiny baby hopping-mouse, eyes still closed, struggles across the accumulation of dry spinifex seeds on the floor of the nest.

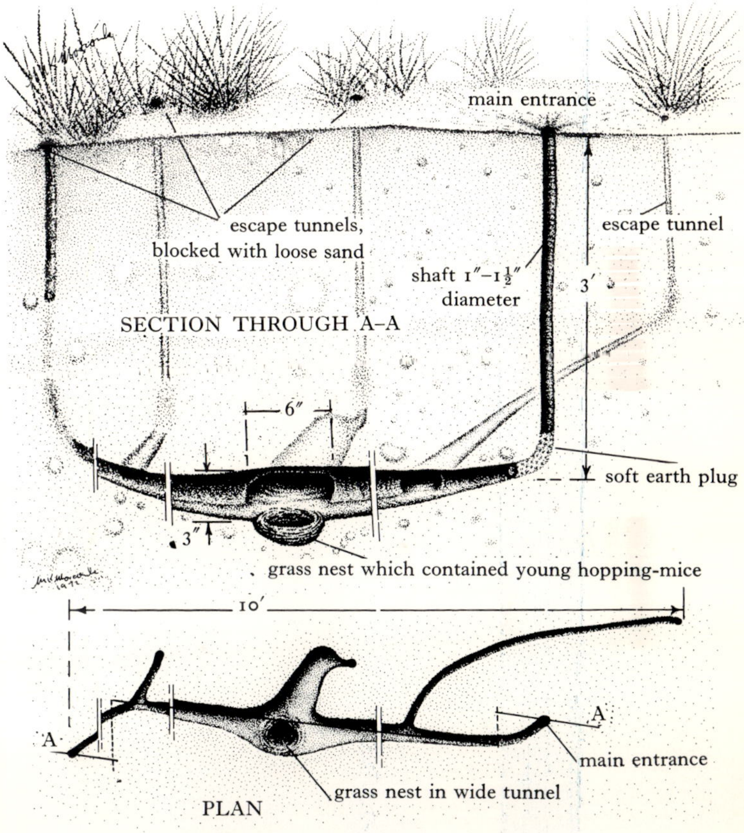

Nest tunnel of the Spinifex Hopping-mouse, *Notomys alexis*

Top Left & Right

The female hopping-mouse about to leave the nest and climb the three-foot vertical shaft to the surface; the young, now a few days older, have more fur.

While the young ones struggle beneath her, the hopping-mouse goes through the important routine of cleaning, washing a foot and leg.

Lower Right

Returning from her nocturnal wanderings, the hopping-mouse dives down the burrow to the young which are now about three weeks old and fully furred.

Now several weeks old, the three very active young mice almost lift their mother from the ground as they struggle beneath her. One of the young hopping-mice, now almost fully grown, studies its mother's beautiful long brush-tipped tail as if contemplating the possible consequences of a sudden sharp nip.

5

Left

Although the Dingo is usually a tawny-yellow dog, a great range of coat colours occurs. This is not the result of crossing with ordinary dogs (which may occur only very rarely in the wild) because the early explorers recorded yellowish-white, blackish and brindled dingos in remote parts of the continent.

Below

Although the Dingo is very closely related to the ordinary dog, it has many distinctive features not seen in other dogs. The ears remain constantly erect, it has a brushy tail, utters yelping and howling sounds instead of barking, and has its own particular gait, and mating behaviour. Combined, these small differences add up to a type of dog not matched by any other. Compared with the marsupials and monotremes, the dingo is a comparative newcomer to Australia, probably being brought here by Aborigines or other wandering native people from the north, thousands of years ago.

Over

The bull sea lion may grow to twelve feet. Male *Neoplaca cinerea* have massive necks and pale yellowish manes. During the breeding season each bull defends his harem of 'clapmatches'.

Right

The female sea lions come ashore during the breeding season, October to December, and give birth to a single pup which is unable to swim until taught. Whenever the female is absent the pup hides in cavities at the first sign of danger. Its fur looks much darker when wet.

Below & Far Right

In defence of her pup the female will rush towards an intruder snarling. When the bull is near he protects the females and pups of his harem. Sea lions inhabit coastal islands from south-western Australia to Kangaroo Island; the only mainland colony is at Point Labatt.

Left
Australia's four species of flying fox can easily distinguished from other bats by their large siz and rather dog-like faces. Although having a somewhat foxy appearance, they are unrelated the canine fox (except that both are mammals) and are perhaps better described by the alternative name, fruit-bat. These giant bats liv in colonies, or camps, often containing tens of thousands of individuals and usually located in swamp, river side or rainforest trees, where the animals hang upside down throughout the day. night the flying foxes spread far and wide in search of native and cultivated fruits, and necta bearing blossoms. This species, the Grey-head Flying-fox, *Pteropus poliocephalus*, inhabits coastal Queensland, New South Wales and Victo

Below
The Little Bat, *Eptesicus pumilus*, flying from th entrance of a Fairy Martin's disused mud nest, found throughout Australia; it also hides in hollows of trees, and caves. It is one of a great many species of small insectivorous bats which swarm through Australian skies at night.

THE MARSUPIAL CARNIVORES

FAMILY DASYURIDAE

Marsupial-mice, native-cats, Tasmanian Devil, Tasmanian Tiger, Numbat

The Broad-footed Marsupial-mice,
genus *Antechinus* and genus *Planigale*

These marsupial-mice have short broad feet, short-haired tapering tails which are never crested or brush-tipped. Insects form the principal part of the diet, so they have a full set of pointed incisor teeth typical of the flesh-eater.

Antechinus apicalis, (formerly *Parantechinus apicalis*)
Dibbler or Freckled Marsupial-mouse

Recognition: The most striking feature of this species is the clear white ring around each eye, visible many feet away. Distinctive also is the tail, which tapers sharply but uniformly from a wide base to a fine point. The speckled colour of the fur is due to each long hair being tipped white. The forearms are reddish.

Distribution: Formerly south-western Australia from Moore River to south coast, east of Albany, now known only from the latter area.

Notes: This nine inch carnivorous marsupial is one of Australia's rarest mammals—it had not been collected for eighty-three years, and consequently was considered possibly extinct. Quite apart from its rarity, it is also a species of scientific interest because of disagreement about its classification, it being placed in the monotypic genus *Parantechinus* by some zoologists, and later, on new evidence, returned to the genus *Antechinus*.

The Dibbler or Freckled Marsupial-mouse is considered by some zoologists to be a connecting link between the marsupial-mice and the native cats.

This rare species was rediscovered when a female was trapped on flowers of *Banksia attenuata* by the author on the night of 26 January 1967. It was the first specimen since 1884. Together with a male captured two nights later, she was kept for a long period for observation.

This rediscovery of the Dibbler was made quite accidently during an attempt to find a specimen of the Honey Possum, which occurs in that locality. Knowing the Honey Possum to be a nectar feeder, the author constructed cage traps which fitted over the large flowers of the low shrubby coastal banksia trees, with the tripwire to spring the door shut passing over the surface of the flower. But, instead of the Honey Possum, it was the far more exciting and rare *Antechinus apicalis* that entered the cage that night.

A third Dibbler was captured soon afterwards, another female, which subsequently gave birth to seven small joeys. All adults and the young were later forwarded to the La Trobe University in Melbourne for studies of behaviour and reproduction that could assist in the conservation of the species. But the adults died within a year and the young reached maturity and eventually died without reproducing.

Since 1967 further attempts by other naturalists to locate live specimens at the original site have been unsuccessful; the Dibbler remains one of the rarest of our mammals, with only three being caught in eighty-eight years, and all from the one half-acre section of low dense scrub on the lower slopes of Mt Manypeaks on the south coast of Western Australia.

From this rediscovery and subsequent studies much new information was obtained about the way of life of the Dibbler. Because the first two were captured on banksia flowers well above the ground, this suggested that they were at least partly arboreal, and that they depend to some extent upon the banksias and other wildflowers for nectar and insects. It appears that the banksias not only provide water (as nectar) during the long rainless summer, but also attract small insects and larger predatory insects and spiders which, in turn, would be an added attraction at the flowers for the carnivorous Dibbler.

This marsupial-mouse can climb quickly and surely, and commonly springs from branch to branch. If suddenly disturbed, it will immediately release its grip and drop two or three feet from the top of the shrub into the very dense ground cover and push under the accumulated debris.

When an insect is found it is first grabbed in the forepaws and held while eaten, the first quick bites being always directed at the head. Large spiders and other potentially dangerous prey are not immediately grasped, but dealt several quick buffeting blows before being snatched up and given a quick disabling bite.

It appears that the Dibbler may not be entirely nocturnal, but moves around beneath the shelter of the very dense low coastal scrub after sunrise, and towards evening. By continuous use of the one hiding place in the matted accumulation of leaves and twigs a rough tunnel is formed, with a chamber at the end.

Yellow-footed Marsupial-mouse,

Related Species:
Antechinus flavipes Yellow-footed Marsupial-mouse or Mardo
Smaller than the Dibbler, fur not speckled, very inconspicuous eye ring. Eastern, south-eastern and south-western Australia, in rainforest, sclerophyll forests and woodlands.

Antechinus stuartii Brown Antechinus, Stuart's Antechinus
Eastern Queensland, New South Wales and Victoria, in rainforest, sclerophyll forests, woodlands and sandstone cave country.

Antechinus godmani Godman's Antechinus
North-eastern Queensland, in mountainous rainforests.

Antechinus bellus Fawn Marsupial-mouse
'Top end' of Northern Territory and Arnhem Land, in tropical savannah-woodlands.

Antechinus rosamondae Little Red Antechinus
Pilbara region of north-western Australia, in spinifex grasslands.

Antechinus swainsonii Dusky Antechinus, Dusky Marsupial-mouse
Coastal New South Wales, Victoria and coastal eastern South Australia, in rainforest and wet sclerophyll forests, alpine heaths and woodlands.

Antechinus minimus Swamp Antechinus, Little Tasmanian Marsupial-mouse
Coastal Victoria to south-eastern South Australia, Tasmania and Bass Strait islands.

Antechinus macdonnellensis Fat-tailed Marsupial-mouse
Northern Australia (Northern Territory and Kimberleys) and Central Australia, arid rocky country.

Antechinus maculatus Pigmy Marsupial-mouse
All mainland States except Victoria.

GENUS *Planigale* Pigmy Antechinuses
The three species of *Planigale* are the smallest known species of marsupial, much smaller than the ordinary mouse, with a head and body length about two and a half inches, no bigger than a large grasshopper, so that quite a prolonged battle ensues when one of the tiny marsupials tackles one of these strong insects. Their heads are rather flattened, triangular from above, helping them to hide in narrow crevices.

Planigale ingrami Flat-headed Marsupial-mouse
Eastern and northern Queensland and extending westward

through the Northern Territory to the Kimberleys and north-west, in tropical savannah woodlands and grasslands.

Planigale tenuirostris Narrow-nosed Planigale
Inland southern Queensland and northern New South Wales, through central Australia to the interior of Western Australia.

Planigale subtilissima Kimberley Planigale
Savannah woodlands and grasslands of the Kimberleys.

The Dunnarts or Narrow-footed Marsupial-mice, GENUS *Sminthopsis*

Most species are about the same size or slightly larger than the ordinary house mouse. They can be separated from the preceding genera by the appearance of the hind feet, which are long and very narrow. The ears are much larger, and the pouch better developed than in the broad-footed marsupial-mice. These are active hunters, attacking and devouring insects and other small creatures at night.

Sminthopsis crassicaudata Fat-tailed Dunnart, Fat-tailed Marsupial-mouse

Recognition: The tail, when fattened, is an unmistakable feature of this and the following four species. Occasionally, however, specimens are caught with thin tails, possibly due to a food shortage. The tail of this species is shorter than head plus body length, the hind feet are narrow, and there are dark patches on the ears and a dark triangle on the head. Unlike ordinary mice, the teeth are sharp-pointed, cat-like.

Distribution: Eastern Australia from south-eastern Queensland, western New South Wales, western Victoria, eastern South Australia, and the south-west of Western Australia.

Notes: Of the many species of *Sminthopsis*, the Fat-tailed Dunnart has been most closely studied. The fattened tail appears to be a food storage, necessary because all fat-tailed species inhabit inland areas varying from moderately dry to extremely arid. The tail becomes fat and spindle-shaped in good seasons when food is abundant, but very thin at times of drought.

All species of *Sminthopsis* are active little predators which will attack, like savage miniature native-cats, any small living creature, usually grasshoppers and other insects, but also small lizards, centipedes, and even house-mice.

The Fat-tailed Dunnart nests in hollow logs and fence posts, or under stones, stumps or clumps of dense vegetation. In its nest the Dunnart, at times of food shortage, becomes torpid thereby saving energy and consequently needing less food.

Fat-tailed Dunnart

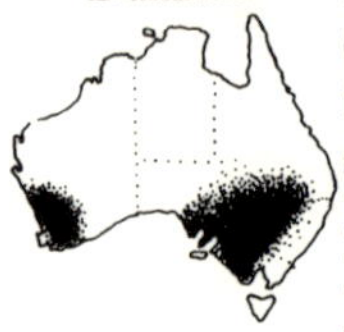

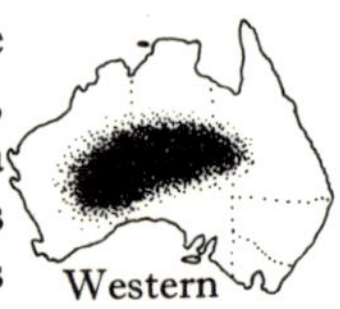

Western Jerboa-marsupial

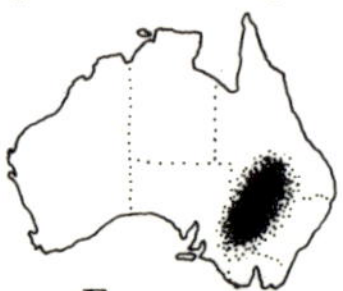

Eastern Jerboa-marsupial

Related Species:
Usually with fattened tails:

Sminthopsis macroura Darling Downs Dunnart
Inland southern Queensland and northern New South Wales, open plains and woodlands.

Sminthopsis hirtipes Hairy-footed Dunnart
Central-Australian desert country.

Sminthopsis granulipes White-tailed or Ashy-grey Dunnart
South-western Australia, extending northwards of the Albany district, in heathlands and woodlands.

Sminthopsis froggatti Stripe-faced Dunnart or Larapinta
From coastal north-western Australia and the southern Kimberleys through central Australia and northern South Australia to western Queensland, in deserts, grasslands and savannah-woodlands. Thin-tailed species.

Sminthopsis murina Mouse-like Sminthopsis, Common Marsupial-mouse
Eastern and south-eastern Australia from Queensland to South Australia, and south-western Australia, in eucalypt forests, swamps, and margins of rainforests.

Sminthopsis leucopus White-footed Dunnart
From eastern New South Wales to southern Victoria and Tasmania, in eucalypt forests.

Brush-tailed Phascogale

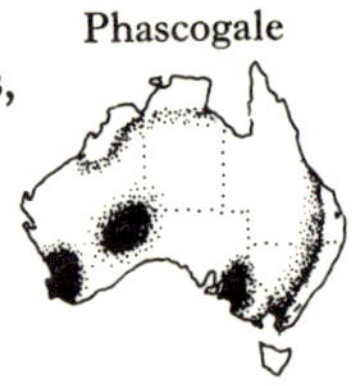

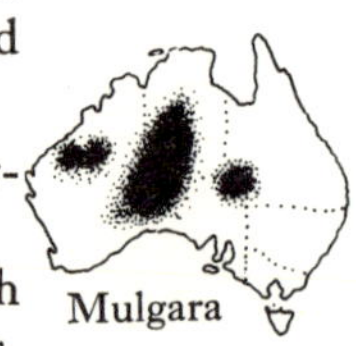

Mulgara

Sminthopsis rufigenis Red-cheeked Dunnart
Tropical northern Australia from north-eastern Queensland to the Kimberleys, in rocky areas and open forests.

Sminthopsis longicaudata Long-tailed Dunnart
Interior of Western Australia to Central Australia.

Sminthopsis nitela Daly River Dunnart
Northern Territory and north Kimberleys, in tropical savannah woodlands.

Sminthopsis psammophila Sandhill Dunnart
Near Lake Amadeus, south-western corner of the Northern Territory. Spinifex-covered sandhills.

The Jerboa Marsupial-mice GENUS *Antechinomys*

Having a superficial resemblance to the hopping-mice *Notomys* (which are not marsupials) these long-legged, long-tailed little insectivorous marsupials were for a long time thought to have a similar hopping way of travelling. But recent studies by Dr W. D. L. Ride, using high-speed photography, and Mr Basil Marlow, recording footprints, showed that their movement is a fast succession of leap-frog actions, with fore-paws touching the ground for an instant ahead of the hind feet producing a 'graceful gallop'.

Antechinomys spenceri Wuhl-wuhl, Pitchi-pitchi, Western Jerboa-marsupial
Desert grasslands, spinifex, saltbush, steppe, woodlands, through Central Australia from western Queensland to the interior of Western Australia.

Antechinomys laniger Kultarr, Eastern Jerboa-marsupial
South-western Queensland through western New South Wales to north-western Victoria.

Tuans, Wambengers or Phascogales GENUS *Phascogale*

The two species of *Phascogale*, considerably larger than the marsupial-mice but smaller than the native-cats, have remarkable bushy brush-tails. Both species are arboreal, nocturnal carnivores, preying upon small vertebrates such as birds and rodents, and large insects. They hide by day in hollow limbs. Length of head and body is nine inches, tail eight inches. The red-tailed species has a rufous colour at the base of the tail between brush and body, where the other species is blue-grey.

Phascogale tapoatafa Brush-tailed Phascogale, Wambenger, Tuan
Eastern Australia from Queensland to Victoria, tropical Northern Territory, and south-western Australia, in rain-forests, eucalypt forests, woodlands and savannah-woodlands.

Phascogale calura Red-tailed Phascogale, Red-tailed Wambenger
Inland parts of Western Australia, Central Australia, and the Murray districts of South Australia, south-western New South Wales, and north-western Victoria, in eucalypt woodlands.

GENUS *Dasycercus*

Dasycercus cristicauda Crest-tailed Marsupial-mouse, Mulgara
The Mulgara superficially resembles the Dibbler in size and shape, but does not have the white eye-ring or speckled fur; the tail is crested instead of tapering. This crest is formed by a brush of long hairs, glossy black, on the upper surface of the terminal half.

Inhabits arid central regions of the Northern Territory, South Australia and Western Australia, in spinifex, rocky and sandridge country.

GENUS *Dasyuroides*

Dasyuroides byrnei Kowari, Byrne's Marsupial-mouse

Considerably larger than the Dibbler, with a total length of about twelve inches, the Kowari is more like the Brush-tailed Phascogale, its long tail having a terminal brush of bristly black hairs.

The Kowari inhabits Central Australia, the Simpson Desert region, in desert steppe and grasslands.

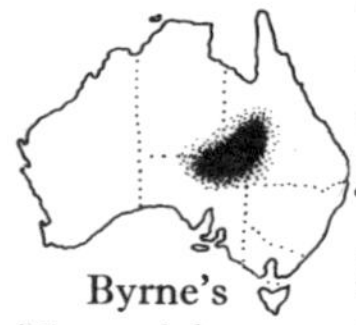

Byrne's Marsupial-mouse

The Native-cats

GENUS *Dasyurus*

The four species of native-cats are nocturnal arboreal predators inhabiting forests and woodlands, at least one species occurring in most parts of Australia. The three small native-cats are spotted except on the tail, while the very much bigger Tiger Cat has spotted body and tail.

Dasyurus maculatus (formerly *Dasyurops maculatus*) Tiger Cat

Tiger Cat

Recognition: Largest of the marsupial carnivores on the Australian mainland, powerfully built, somewhat larger than the biggest of domestic cats, with a body length commonly reaching twenty-five inches and tail nineteen inches. The muzzle is blunt compared with the sharp-faced little native-cats, and the powerful jaws open extremely wide, well in excess of ninety degrees. The ears are short and rounded. The Tiger Cat differs from the Eastern Native-cat, the only similar marsupial occurring in the major part of the Tiger Cat's range, in having five toes on the hind feet and a spotted tail, while the small species has four toes on the hind feet and an unspotted tail.

Distribution: Rainforests and eucalypt forests, eastern Queensland and New South Wales in southern Victoria, the south-east of South Australia, and Tasmania.

Notes: The largest of the arboreal carnivores, and exceeded in size only by the Tasmanian Tiger and Tasmanian Devil, Tiger Cats are able to take quite large prey, including the smaller wallabies and rat-kangaroos—one was said to have killed a large tomcat after a terrific battle. Like other marsupial carnivores, the Tiger Cat shows a complete lack of fear in the presence of food and in the past, when Tiger Cats were common, individuals that raided domestic fowl pens at night were so bold and returned so persistently that trapping or shooting became not only necessary but almost inevitable.

Tiger Cats are skilled climbers, spending most of their time in trees, and showing a preference for the heavy forests and dense scrubs of coasts and mountain ranges. On the palms of front and hind feet are serrated pads which give a non-slip grip on the branches, and the claws are long and sharp.

Four to six young are carried in a pouch which is a flap of skin covering the front and sides of the mammary area. In Victoria the season of birth is said to be May.

The Tiger Cat has become uncommon over most of its mainland range except in some of the wilder and more remote areas. The specimen photographed was in rainforest at an altitude of about 3,000 feet on the mountainous eastern slopes of the Great Dividing Range, north-eastern New South Wales. The species is more common in Tasmania.

Western Native-cat, Chudich

Dasyurus geoffroii (formerly *Dasyurinus geoffroii*)

Western Native-cat

Recognition: At first glance, a smaller edition of the Tiger Cat, but differs in having a sharper-pointed face, ears relatively larger and more pointed, tail unspotted, and the pads on the soles of its feet are granular-surfaced instead of ridged.

Distribution: Formerly very widespread, occurring in western New South Wales, Queensland except the northern half and the south-east, north-western Victoria, South Australia except the south-east, Central Australia, and Western Australia except the Kimberleys. This species is probably now extinct in eastern Australia, but quite common in forested south-western Australia where there occurs a larger sub-species, *Dasyurus geoffroii fortis*.

Notes: Native-cats are of zoological interest as one of the few examples of superfetation among Australian marsupials. This is the over-production of young, there being as many as eighteen born on occasions, but the pouch contains places for eight only. The first (probably the strongest and first-born) to reach the pouch and attach to a teat can be the only ones to survive. After leaving the pouch, the young clamber all over the mother as she hunts at night, eventually almost dragging her to the ground.

Although native-cats attack and kill poultry, they are very beneficial, feeding upon mice and destructive insects.

Related Species:

Dasyurus viverrinus (formerly *Dasyurus quoll*) Eastern Native-cat, Quoll

Forests along coasts and mountain ranges of eastern and south-eastern Australia from northern New South Wales to South Australia and Tasmania.

Dasyurus hallucatus (formerly *Satanellus hallucatus*) Little Northern Native-cat, Satanellus

Tropical Australia, principally coastal, from north-eastern Queensland to the Kimberleys and north-west.

GENUS *Sarcophilus*

Sarcophilus harrisii Tasmanian Devil

Tasmanian Devil

Identification: Large, very solidly and powerfully built, entirely black except for white bands usually present across chest and rump. Length of head and body twenty-eight inches, tail twelve inches. Muzzle short and broad, ears short and rounded.

Distribution: Now confined to Tasmania, but occurred on the mainland before the coming of European man. Fossil remains have been found in eastern, south-eastern and south-western Australia.

Notes: Among the marsupial carnivores, the Devil is exceeded in size only by the Tasmanian Tiger. It is principally terrestrial (but able to climb) and nocturnal in habits. Its prey includes small mammals, reptiles and any carrion it can find. Sheep that are weak or penned are liable to be attacked. The Devil is a most efficient scavenger, cleaning up every part of a carcass, including most of the bones. Tasmanian Devils are found mainly in sclerophyll forest, where they shelter in hollow logs or rock crevices.

Four young are born in early winter, carried at first in the pouch, and later clinging on their mother's back.

GENUS *Thylacinus*

Thylacinus cyanocephalus Tasmanian Wolf, Tasmanian Tiger, Thylacine

Tasmanian Wolf

Recognition: The largest marsupial carnivore, wolf-like in shape and size, but with hind quarters tapering to a heavy kangaroo-like tail. Across the lower back and rump are sixteen dark transverse stripes. Head and body length forty-four inches, tail twenty-one inches.

Distribution: Confined to Tasmania, but did occur on the mainland prior to the coming of European man. Fossil remains have been found in south-eastern and south-western Australia.

Notes: A mummified carcass of a Thylacine found in a cave at the western edge of the Nullarbor still had fur showing the characteristic stripes. Radio-carbon dating showed that the animal had died between two and three thousand years B.C., so the species must have persisted on the mainland until comparatively recently. In Tasmania the Thylacine was hunted almost to extinction, but it is thought that a few may still survive in the remote south-west of the island.

The home and refuge of the Numbat is a hollow log.

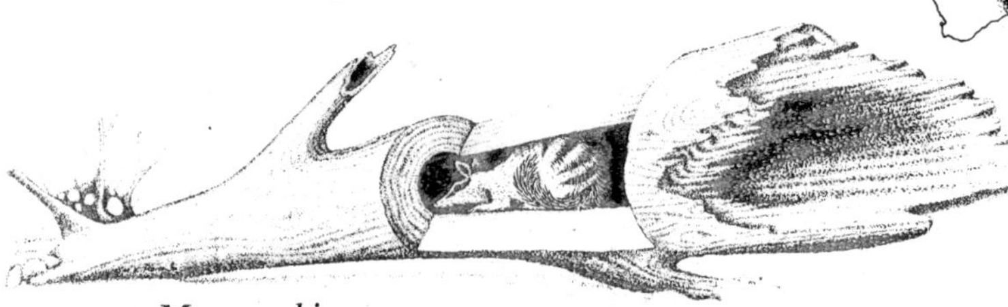

GENUS *Myrmecobius*

Myrmecobius fasciatus Numbat, Banded Anteater, Marsupial Anteater

Recognition: A slender animal with long bushy tail and pointed snout. There are about six very conspicuous white stripes across the dark red-brown lower back and rump, and a dark stripe through the eye. The head and body length is about ten inches, tail seven inches.

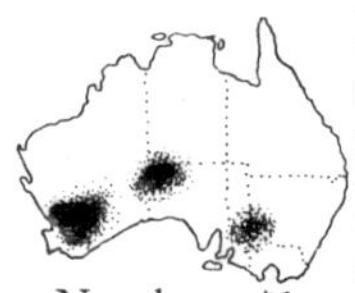

Numbat

Distribution: There are two sub-species of Banded Anteater: *Myrmecobius fasciatus fasciatus* inhabiting woodlands of south-western Australia, and *Myrmecobius fasciatus rufus* the desert areas of far eastern Western Australia, north-western South Australia (including the Everard Ranges). The latter sub-species once occurred as far east as south-western New South Wales.

Notes: The Numbat is an aberrant member of the native-cat family, Dasyuridae, but appears so different that it has sometimes been placed in a family of its own, the Myrmecobiidae. However it has many characteristics in common with the dasyures—certain patterns of behaviour are dasyure-like; the molar teeth, although degenerate due to the soft termite diet, closely resemble dasyurid teeth; and the number and morphology of the chromosomes is identical to those of all other members of the family Dasyuridae.

The Numbat is very closely tied to the ecology of natural Wandoo woodland, where the trees are scattered, the forest floor partly open and partly covered with low scrub. Most of the mature Wandoo trees have been eaten hollow by termites, and the ground throughout the area is littered with branches and logs which are almost invariably hollow. The Wandoos and the termites together provide the Numbat with abundant hollows for shelter and protection against such predators as the Goshawk and the Wedgetail Eagle, while the termites are almost its only food.

Unlike the majority of small mammals, Numbats are abroad in full daylight, not at night, and can be observed as they go about their activities.

This is usually a solitary animal which spends most of its time searching for termites, digging into the ground and turning over small pieces of wood. During cold winter months it spends considerable time lying in the sun. If disturbed, the Numbat runs swiftly towards one of the hollow logs in its territory, the bushy tail arched high over its back, perhaps giving some protection against a swooping Goshawk or eagle. Now and then it may stop, stand upright on hind legs, and look around. The Numbat is exceptional among mammals in that a wild one can be handled and, although struggling to escape, will not bite or scratch.

The young are born between January and April. As the Numbat has no pouch the four young (the normal litter) cling to the long hairs of the pouch area under the mother's belly with their hands, but are held even more securely by the teats which, as with all marsupials, swell in their mouths to make a connection that will not be broken for many months. During this time (at least while they are just an inch or two in length) the female pays them no attention, and even seems unaware of their presence. Later they are left in a hollow log or hole in the ground while she is out foraging.

Marsupial-mole

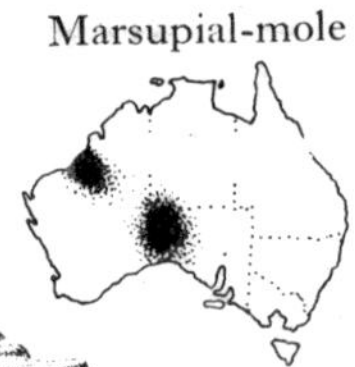

FAMILY NOTORYCTIDAE

GENUS *Notoryctes*

Notoryctes typhlops The Marsupial-mole

Recognition: In appearance resembles moles of other continents, but very small, four to six inches total length. It has no eyes and no external ears. The forefeet are highly modified for digging, with large triangular shovel-like claws on the joined third and fourth toes. The fur is velvety, white to golden orange with an iridescent sheen.

Distribution: Two separate populations are known, one extending from southern Northern Territory through western South Australia to the Bight, the other in the Great Sandy Desert of north-western Australia. Those from the north-west are in some publications listed as a separate species, *N. caurinus*.

Notes: The Marsupial-mole presents a remarkable example of convergent evolution. Although not related to the moles of other continents, the similarity of the subterranean way of life has resulted in animals of two different origins coming to look very much alike. The Marsupial-mole lives in sand-ridge desert country and is rarely seen. It lives on insects, their larvae and pupae, and is said to emerge from the ground only in wet weather.

BANDICOOTS FAMILY PERAMELIDAE

The animals of this family are mixed-feeders, that is, they eat insects, vegetable matter, occasionally small mammals, and lizards. The food preferences vary considerably from one genus to another, some groups being entirely insectivorous, some entirely vegetarian.

These mixed feeding habits are reflected in certain main characteristics by which the bandicoots are classified: they possess the many-incisored or polyprotodont dentition of the marsupial carnivores, but have a syndactyl foot structure the same as the diprotodont vegetarian possums and kangaroos.

Short-nosed Bandicoot

Short-nosed Bandicoots

GENUS *Isoodon*

Although all marsupial bandicoots have long slender snouts the members of this genus are short-nosed by comparison with some other genera.

Isoodon obesulus Quenda, Brown Bandicoot, Southern Short-nosed Bandicoot

Recognition: About the size of a rabbit, length of head and body approximately fourteen inches, and the tail is short and tapering five inches. The muzzle is pointed, the ears short and rounded, fur coarse and grizzled yellowish-brown, white on belly.

Distribution: East-coastal Australia from Cape York Peninsula through New South Wales to southern Victoria and south-eastern South Australia. Also in Tasmania and Western Australia.

Notes: Nocturnal and insectivorous, inhabiting forests and heathlands wherever there is dense ground cover. The presence of bandicoots can be detected by the little conical pits left where they have been digging for beetles and insect larvae.

A rough nest of sticks is constructed on the ground in dense vegetation. There is usually no entrance as the bandicoot just burrows under the pile of debris. Seven young form the usual complete litter.

Related Species:

Isoodon macrourus Brindled Bandicoot
Northern and eastern Australia, from the Kimberleys through Arnhem Land, north Queensland, to north-eastern New South Wales. Eucalypt forests and woodlands where there is good ground cover.

Isoodon auratus (includes *I. barrowensis*) Golden Bandicoot, Wintarro
From north-western Australia and Barrow Island to Central Australia.

The Long-nosed Bandicoots
GENUS *Perameles* and GENUS *Echymipera*
The genus *Perameles* contains many common species. They are nocturnal, insectivorous, and have an incredibly long thin, almost tubular, snout. Compared with the short-nosed bandicoots, they have longer legs and ears, and look more lightly built. All but one species, the Common Long-nosed Bandicoot, have a pattern of bars across the hindquarters, and on that one species the bars are faintly visible on juvenile animals.

The genus *Echymipera* differs from the two preceding genera in the reduction of the upper incisors to four pairs instead of five.

Perameles nasuta Common Long-nosed Bandicoot
East-coastal Australia from Queensland to Victoria, in rain-forest, sclerophyll forests and woodlands.

Perameles eremiana Desert Bandicoot, Orange-backed Bandicoot
Spinifex grasslands of Central Australia, and adjacent parts of Western Australia and South Australia.

Perameles gunnii Tasmanian Barred Bandicoot, Striped Bandicoot
Tasmania and southern Victoria, in woodlands and open country where there is low dense ground cover.

Perameles bougainvillei (includes *P. fasciata*) Marl, Little Marl, Barred Bandicoot
In heaths and dune vegetation from inland New South Wales and north-western Victoria through southern parts of South Australia to sand plain areas of south-western Australia.

Echymipera rufescens Spiny Bandicoot
Confined to the rainforests of the McIlwraith Range, Cape York, north Queensland.

Rabbit-eared Bandicoot, *Macrotis lagotis*

Rabbit-eared Bandicoots
GENUS *Macrotis* (formerly *Thalacomys*)
The two bandicoots of this genus are among the most beautiful and graceful of all native mammals, having large rabbit-like ears, and long silky black-and-white or all white tails. These bandicoots are powerful diggers, and always live in a deep burrow. They are carnivorous, feeding principally upon termites, beetle larvae and other insects.

Common Rabbit-eared Bandicoot

Rabbit-eared bandicoots are now almost extinct except for remote arid parts of the far north and interior.

Macrotis lagotis Dalgyte, Bilby, Common Rabbit-eared Bandicoot
Length head and body eighteen inches, tail nine inches. Formerly occurred in Western Australia from the Kimberleys to south-west, western and northern South Australia, western New South Wales, and far south-western Queensland.

Macrotis leucura Yallara, Lesser Rabbit-eared Bandicoot
Tail all white; smaller, head and body ten inches.
Sandhills of Simpson Desert region, Central Australia.

Pig-footed Bandicoots
GENUS *Chaeropus*
Somewhat smaller than a rabbit, the one species of this genus looked rather like a miniature deer with long ears, slightly crested tail, and long thin legs. On each hind foot is one large functional toe, and on the forefeet only two functional toes which together resemble the cloven feet of a pig. This marsupial has become extremely rare, possibly extinct. The last specimen was collected in 1907, and the last reports of sightings, in 1926, came from the Aborigines of the Musgrave Ranges. It seems to have been principally vegetarian.

Chaerops ecaudatus Pig-footed Bandicoot
Arid and semi-arid woodlands, mallee scrub and grasslands of Central Australia, inland Western Australia, South Australia except the south-east, south-western New South Wales, and north-western Victoria

THE HERBIVOROUS MARSUPIALS
DIPROTODONTA

THE KOALA FAMILY PHASCOLARCTIDAE

The single species of this family is one of the most attractive and interesting of mammals. It is one of the most highly specialised of herbivorous marsupials, and seems more closely related to the wombats than to any other group, but at the same time, not very closely related, as shown by its placement in a separate family.

GENUS *Phascolarctos*
Phascolarctos cinereus Koala
Recognition: Large (head and body thirty-two inches), tail-less, snout bulbous and naked of fur, first and second toes of the forefoot opposable to the other three toes.

Distribution: Eastern coastal Australia from Townsville to Melbourne (formerly to South Australia), extending inland to the western slopes of the Great Dividing Range, in eucalypt forests and woodlands.

Notes: The Koala is not only specialised for an arboreal life, but also in its diet of nothing but eucalyptus leaves. To accommodate this bulky food its 'appendix', an additional prolongation of the intestines, is six to eight feet in length. The Koala chooses its leaves with great care; only the leaves of about a dozen different trees are acceptable, but at certain times of the year the young leaves of some species are rejected because they contain poisonous substances that can be fatal if eaten.

The Koala gives birth to a single, very small, young (rarely twins) which, after the usual period of growth in the backwards-opening pouch, rides on its mother's back. Koalas are very slow breeding, probably producing one young per year, and this has no doubt contributed to their widespread decline in numbers following trapping for skins, introduction of diseases, increased frequency of bushfires, and extensive land clearing.

THE WOMBATS FAMILY VOMBATIDAE (formerly PHASCOLOMIDAE)

Although wombats show signs of once having been arboreal marsupials, sharing some distant past ancestry with possums and the koala, they have since had a very long period as terrestrial animals, becoming so specialised for their present way of life that they must be placed in a family of their own. A distinctive feature is their rodent-like dentition which is unlike that of any other marsupial. There is only a single pair of upper and lower incisors which, like the incisors of rodents, grow continuously and are ground away and sharpened in use. Wombats are entirely vegetarian.

All wombats are powerful diggers, their stocky build and shovel-like claws enabling them to excavate burrows up to a hundred feet in length.

Their well-formed pouch, like that of the koala, opens towards the rear.

Forest Wombats

GENUS *Vombatus*

The snout is naked of fur and its skin granular in texture; the body fur is thick and coarse and the ears short.

Vombatus ursinus (formerly *Vombatus hirsutus* or *Phascolomis mitchelli*)

In this species are included three slightly differing sub-species: the Common Wombat of mainland south-eastern Australia, the Island Wombat of the Bass Strait islands, and the Tasmanian Wombat.

Brushtail Possum

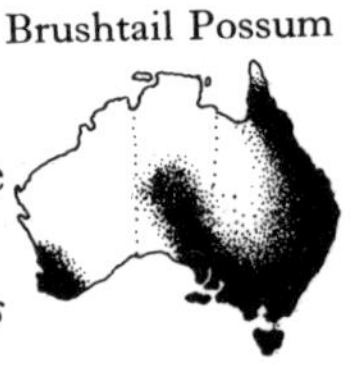

Common Wombat

Plains Wombats

GENUS *Lasiorhinus*

The tip of the snout has fine hairs and differs in shape to the noses of the preceding genus, the ears are much longer and pointed, and the body fur is soft and silky.

Lasiorhinus latifrons Hairy-nosed Wombat

Open woodlands, grasslands and semi-arid shrub steppe, from the extreme south-east of Western Australia through southern South Australia to the Murray River.

Hairy-nosed Wombat

Lasiorhinus barnardi Queensland Hairy-nosed Wombat

East-central Queensland and Riverina district of New South Wales.

Lasiorhinus gillespiei Moonie River Wombat

South-eastern Queensland.

FAMILY PHALANGERIDAE

The cuscuses and large possums are members of the related genera *Phalanger*, *Wyulda* and *Trichosurus*. Unlike the Koala, wombats and bandicoots, these have forward-opening pouches.

The Cuscuses

GENUS *Phalanger*

Australia has two species of cuscuses, both confined to Cape York Peninsula. The true home of this genus is New Guinea and adjacent islands where many species occur.

Phalanger maculatus (formerly *Spilocuscus nudicaudatus*) Spotted Cuscus, Spotted Phalanger

Spotted Cuscus

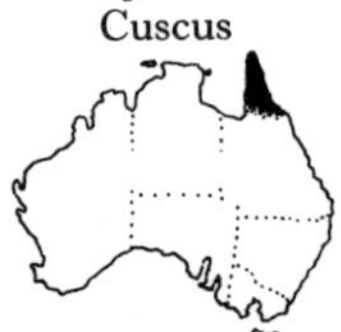

Recognition: These rather monkey-like possums are distinguished by their tails which are devoid of fur on the terminal half, and have rasp-like scales right along the undersurface. The ears are extremely small, scarcely showing above the fur. Colour is variable, grey to rusty-brown, spotted or plain, sometimes almost white. Length, head and body twenty-six inches, tail nineteen inches.

Distribution: Rainforests and gallery forests of north-eastern Cape York, extending as far south as the McIlwraith Range.

Notes: A slow-moving nocturnal arboreal marsupial, living principally upon leaves and fruit, but including also in its diet small mammals and birds. Litter size ranges from two to four.

Related Species:

Phalanger orientalis Grey Cuscus

Resembles the preceding species, but on both sexes there is a dark dorsal stripe extending from between the ears to the rump. Inhabits rainforests and gallery forests of Cape York, north of the McIlwraith Range.

GENUS *Wyulda*

The one species of this genus has a remarkable prehensile tail which is covered for almost all of its length on all sides with rasp-like scales instead of the usual fur.

Wyulda squamicaudata Scaly-tailed Possum

Inhabits the trees of rocky country of the western, northern and central Kimberley region of Western Australia.

The Brushtail Possums GENUS *Trichosurus*

These large possums have a bushy tail which is naked of fur along its undersurface on the terminal half, longer ears than other possums, and a rather pointed, foxy face. The teeth indicate mixed feeding rather than a purely vegetarian diet.

Trichosurus vulpeca Brushtail Possum, Brush Possum, Common Possum

Recognition: Silvery-grey, sometimes almost black, yellowish beneath. Tail bushy, black at the tip, often white-tipped in Western Australia. Head and body eighteen inches, tail eleven inches.

Distribution: Very widespread, with many geographical races; occurs in Tasmania, south-eastern, eastern, northern, south-western and parts of the interior of Australia.

Notes: One of the most successful of marsupials, the Brushtail has maintained its numbers in spite of man's changes to the environment. It shows its adaptability by taking up residence in house roofs instead of tree hollows. In regions where hollows are scarce, it lives in rabbit burrows or rock crevices.

Among about seven sub-species are the Coppery Brushtail (*Trichosurus vulpeca johnstoni*), the south-western Brushtail (*T. v. hypoleucus*), and the Tasmanian Brushtail (*T. v. fuliginosus*), all well-defined in colour, markings and distribution.

Related Species:

Trichosurus caninus Short-eared Brushtail, Mountain Possum

Occurs in eastern Australia, from southern Queensland to Victoria, in heavily forested coastal mountain ranges.

Trichosurus arnhemensis (formerly *T. vulpeca arnhemensis*) Northern Brushtail Possum

'Top End' of the Northern Territory, Kimberley and Barrow Island.

FAMILY PETAURIDAE

The ringtail possums, the gliders, the Bushy-tailed Ringtail and Leadbeater's Possum.

The Ringtail Possums, GENUS *Pseudocheirus*, GENUS *Hemibelideus*, and GENUS *Petropseudes*

Slow-moving possums distinguished by their very long tapering tails which are carried curled in a ring when not being used to grip branches. The ears are very short. The first and second toes of the forefeet are opposable to the other three toes for grasping.

Pseudocheirus peregrinus Common Ringtail
Recognition: Head and body usually rufous-grey, grey-white beneath, white tip to the tail, white behind the ears. (Considerable variation in different regions.)

Distribution: From north Queensland through New South Wales to Victoria, Tasmania, South Australia and south-western Australia.

Notes: These nocturnal, arboreal marsupials sometimes live in hollows of trees, but more often build large globular nests in dense vegetation. Ringtails are entirely herbivorous, feeding principally on leaves and fruits.

A number of populations of this species, isolated in different parts of Australia, differ sufficiently in appearance to have been described as separate species in the past. Recent studies of their blood proteins by Dr J. Kirsch have shown them to be no more than local races of the Common Ringtail.

These sub-species include the Tasmanian Ringtail (*Pseudocheirus peregrinus convolutor*), the Western Ringtail (*P. p. occidentalis*), and the Bunya or Rufous Ringtail (*P. p. rubidus*). In all there are about eleven different sub-species.

Related Species:
Pseudocheirus archeri Green Ringtail, Striped Ringtail, Toolah
Greenish-brown fur, two stripes down the back. Inhabits rainforests of the Atherton Tableland area of north Queensland.

Pseudocheirus herbertensis Herbert River Ringtail, Mongan
Chocolate-brown fur and pure white undersurface. Rainforests of north-east Queensland.

Hemibelideus lemuroides Lemur-like Ringtail, Bushy-tailed Ringtail
Resembles preceding species, but with bushy-furred prehensile tail. Fur dark chocolate brown. Rainforests of Atherton Tableland area of north Queensland.

Petropseudes dahli Rock-haunting Ringtail, Wogoit
Toes of front feet not opposable; very short tail. Lives among boulders of rocky ranges of the far north from the Kimberleys to western Arnhem Land.

Large Gliders

GENUS *Petaurus* and GENUS *Schoinobates*

These possums glide from tree to tree by means of their flight membranes, an extension of the body skin which is joined to the legs so that it is tightly stretched when the animal leaps from a tree with all four limbs outspread; the tails are long and bushy to steer, like a rudder, in flight.

Petaurus breviceps Sugar Glider
Recognition: Medium-sized, superficially squirrel-like, head and body seven inches, tail eight inches. Fur ash-grey, almost white below, dark dorsal stripe. Gliding membrane edged with black, and extending out to wrists and ankles. Tail grey, with long fluffy fur. Toes of forefeet not opposable.

Distribution: Northern, eastern and south-eastern coastal Australia, from the Kimberleys to south-eastern South Australia, in eucalypt forests and woodlands.

Notes: Sugar Gliders live in small communities or family parties, probably the young of several seasons still living with the parents. The two young remain in the pouch for several months until they become too heavy to be carried by the mother on her nocturnal flights, when they are left in their tree-hollow nest. Sugar Gliders are very active, fast-moving in their nocturnal hunting for insects, nectar fruits, and the sweet gum that exudes from the bark of some eucalypts.

Squirrel Glider

Common Ringtail

Greater Glider

Green Ringtail

Herbert River Ringtail

Sugar Glider

Feathertail Glider

Related Species:
Petaurus norfolcensis Squirrel Glider
Similar to the preceding species but larger, head and body ten inches. Occurs in eucalypt forests and woodlands of eastern and inland Queensland, New South Wales and Victoria.

Petaurus australis Fluffy Glider, Yellow-bellied Glider
Larger than both preceding species, head and body twelve inches, total length seventeen inches. Dark stripe down centre of back, undersurface creamy to orange. Inhabits eucalypt forests of mountainous coastal eastern Australia from Queensland through New South Wales to Victoria.

Schoinobates volans Greater Glider, Dusky Glider
The largest species of glider, head and body seventeen inches, tail twenty inches (northern sub-species *S. v. minor* is smaller, head and body twelve inches). Gliding membrane reaches only to elbows. Coastal forests and woodlands from north Queensland to Victoria.

GENUS *Gymnobelideus*

Gymnobelideus leadbeateri Leadbeater's Possum
A squirrel-like possum with a very long bushy tail. Resembles the Sugar Glider, but without gliding membrane. A species once thought to be extinct, but now known to inhabit dense eucalypt forests of south-eastern and southern Victoria.

GENUS *Dactylopsila*

Dactylopsila trivirgata Striped Possum
A rather small, squirrel-like possum with conspicuous black-and-white stripes down the back and sides. Inhabits the rainforests and adjacent eucalypt forests of north-east Queensland.

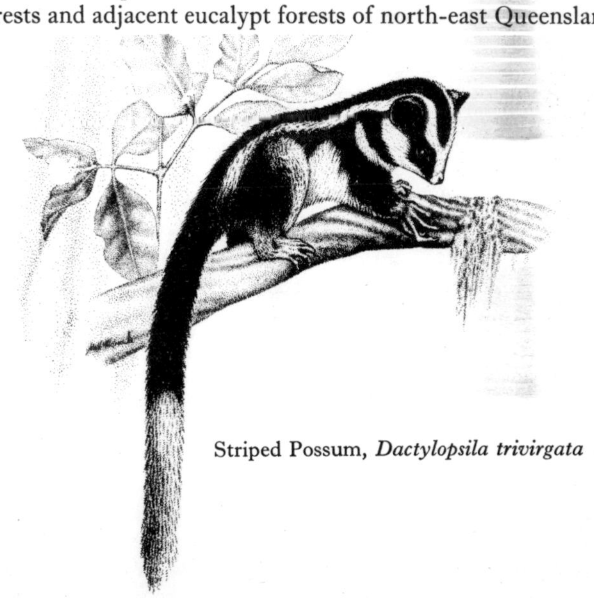
Striped Possum, *Dactylopsila trivirgata*

SMALL POSSUMS AND GLIDERS

FAMILY BURRAMYIDAE

The pigmy possums and pigmy glider are mouse-like in size and superficial appearance, but differ from mice in having pouches to carry their young, many small upper incisor teeth instead of the typical single pair of rodent incisors, and prehensile tails. These possums have syndactyl toes (joined second and third toes on the hind feet with comb-like double claw) which separates them not only from all placental mice but also from the dasyurid marsupial-mice.

GENUS *Acrobates*

Acrobates pygmaeus Feathertail Glider, Pigmy Glider
Recognition: Mouse-sized (head and body three inches, tail three inches), prehensile tail with a row of stiff bristles on each side resembling an emu feather. Flight membrane between wrists and ankles. It is the only really tiny glider.

Distribution: Eastern and south-eastern Australian eucalypt forests from Cape York to Adelaide.

Notes: This, the smallest of gliders, lives in small groups in hollows of trees during the day. A globular nest of gum leaves and shredded bark is constructed and when occupied the entrance is closed over. At night the tiny gliders hunt actively for insects; nectar of flowering trees makes up a large part of the diet. The four young, the maximum normal litter, are carried in a pouch while small.

GENUS *Cercartetus* and GENUS *Burramys*

These tiny pigmy possums are similar to the preceding species, but lacking the flight membranes and the feather-like edging to the tail. The fully prehensile tail is often carried rolled up like that of the ringtail possums.

Cercartetus concinnus South-west Pigmy Possum, Mundarda

Recognition: About the size of a mouse, with long prehensile tail (head and body four inches, tail four and a half inches), fur fawny-brown to grey-brown, underparts paler, or near-white.

Distribution: South-western Australia, south-eastern South Australia, western Victoria and south-western New South Wales.

Notes: This solitary, arboreal, nocturnal marsupial lives upon nectar and insects. During the day it hides in abandoned bird nests, under loose slabs of bark, or in small hollows in trees. Pigmy possums are able to lower their body temperature and become torpid, a state rather similar to hibernation.

Related Species:

Cercartetus nanus Eastern Pigmy Possum

Very similar to the preceding species, belly fur sometimes more grey than creamy. Inhabits sclerophyll forests and woodlands of coastal eastern and south-eastern Australia and Tasmania.

Cercartetus lepidus Tasmanian Pigmy Possum, Little Pigmy Possum

Smaller size, otherwise resembling preceding species. Forests of Tasmania and Kangaroo Island.

Cercartetus caudatus Long-tailed Pigmy Possum

The unusually long tail is about one and a half times the head and body length. North-eastern Queensland, and New Guinea.

Burramys parvus Mountain Pigmy Possum

A species previously known only from fossil remains; the first living specimen was found in 1966 in the Alps of eastern Victoria.

FAMILY TARSIPEDIDAE

This family contains just one species, the Honey Possum. Its body, snout and limbs have become so extremely modified that little is known of its ancestry except that it has no close relatives among the marsupials that inhabit Australia today. It is only distantly related to the similar-sized pigmy possums.

GENUS *Tarsipes*

Tarsipes spenserae Noolbenger, Honey Possum

Recognition: About the size of a mouse, head and body three inches, tail three and a half inches. Has three well-defined stripes down the back, a very long whip-like tapering prehensile tail with a naked area under the tip, and an extremely elongated almost tubular snout.

Distribution: Confined to the south-western corner of Western Australia, from the Murchison River on the west coast to Esperance on the south coast, usually inhabiting sand plains, tree and shrub heathlands.

Notes: The most extreme adaptations for the nectar-eating way of life are to be seen in the modifications of the head, the snout being used like the long beak of the honeyeating birds, to probe into the wildflowers for nectar. Many of the commonest wildflowers of its habitat are deeply tubular (grevilleas, lambertias) or brush-like (banksias, bottlebrushes) so the nectar can be reached only by an animal equipped to reach into such narrow confines.

The Honey Possum's snout has become lengthened until almost a trunk or proboscis. There are flanges along its lips which overlap to form a tubular channel through which nectar, pollen and microscopic insects (living in huge numbers in some flowers) can be sucked up. Its slender tongue is covered with fine bristles, with a tuft at the tip, for licking up nectar, and darts out an inch beyond the tiny animal's nose; the teeth have greatly degenerated after millions of years of such soft food.

South-western Pigmy Possum

Honey Possums live in a region where, on the heathlands, there are at any time of the year some wildflowers in bloom, autumn and winter included. The little marsupials move about, possibly in colonies, according to the flowering, at one season being found among bottlebrush shrubs, another time among the banksias.

During the day Honey Possums hide in nests of grass and fur, usually built in dense foliage, such as the tops of grass-trees; they are occasionally found curled up in the abandoned nests of small birds. Their young are probably kept in this nest for a considerable time after leaving the pouch.

KANGAROOS FAMILY MACROPODIDAE

Members of the macropod family can readily be distinguished from all other marsupials by a number of characteristics. The hind limbs are very large in proportion to the rest of the body, and adapted for a two-footed leaping action. The forelimbs are much shorter and more lightly built. The tail is large, heavy, and used as a balance during fast progression, and as a kind of auxiliary limb when standing or moving slowly in quadrupedal fashion.

The foot is very long (hence the term 'macropod') with its fourth toe very large and heavily clawed. The other toes are smaller, and the second and third toes very small, united except for the claws. These twin-clawed joined pairs of toes on each hind foot are termed 'syndactylous', and are used like a comb in grooming the fur.

The macropods (except for one species, *Hypsiprymnodon*) have evolved a digestive system parallel to that of the ruminants of other continents, the stomach being large and sacculated, so that bacteria and protozoa contained in large numbers can carry out a preliminary digestion of the bulky foliage or herbage food. The teeth are modified for grazing, with two very large scissor-like incisors at the lower front which press up against a leathery pad between the semicircle of the upper incisors.

Honey Possum

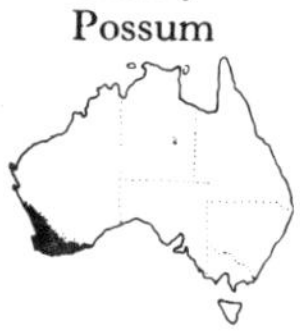

The larger macropods are known as kangaroos, the smaller ones as wallabies, but there is no distinct difference between kangaroos and wallabies other than size, and even in this respect the demarcation is not clear, there being little difference between the smallest kangaroos (such as the Barrow Island race of the Euro), and the largest wallabies.

Large Kangaroos

GENUS *Megaleia* and GENUS *Macropus*

Within these two genera are three groups of large kangaroos—(i) the Red Kangaroo; (ii) the two species and numerous subspecies of grey kangaroos; (iii) and the Wallaroo and Euro group.

The three groups can be separated on a number of morphological characters. The Red Kangaroo has very short dense woolly fur, the Wallaroo and Euro have long thick shaggy fur, while the greys are more or less intermediate between those extremes. Similarly, the Red Kangaroo has the longest hind limbs and the Wallaroo and Euro the shortest—contributing to the stocky appearance of the latter group.

The noses also separate the three groups. The Wallaroo and Euro have a large bare area of black skin around and especially above the nostrils; the grey kangaroos have the nose almost entirely covered with fur except for a very narrow band of bare skin above each nostril; the Red Kangaroo is intermediate.

Megaleia rufa Red Kangaroo, Plains Kangaroo
Recognition: Very large, head and body sixty-five inches, tail forty-two inches, weight of males up to 180 pounds. Colour varies in different regions and may be entirely reddish, red-and-grey, or entirely greyish, while females are generally of a different colour to the males of their locality. In addition to preceding notes on differences between groups of large kangaroos, the Red Kangaroo always has distinct white and black markings each side of the muzzle. Its tail becomes paler towards the tip.

Distribution: Very widespread across the whole continent except for the far north, the forested and mountainous east-coastal and south-east, and the forested south-west.

Notes: More is known of the biology of the Red than of any other large kangaroo due to the work of Dr H. J. Frith and his colleagues of the C.S.I.R.O. Division of Wildlife Research. These studies were necessary for the effective conservation of this species which has been extensively exploited for pet food, becoming the basis of a very large industry. In addition, the large kangaroos have always been under pressure from graziers who believed that they compete with sheep for available pasture.

The research has revealed that the kangaroo normally does not select the same food plants as sheep and, weight for weight, kangaroos do not eat more than sheep. However kangaroos do convert grass into meat far more efficiently, a kangaroo being 52% meat, a sheep 27%.

Kangaroo numbers must be controlled and sometimes reduced, so a well regulated industry based on sound scientific data could give a permanently sustained yield. But in some States the indiscriminate, uncontrolled and very wasteful slaughter could result in the kangaroo meat industry destroying both itself and the natural resource which it is exploiting.

The Red Kangaroo is principally an inhabitant of the semi-arid plains of the interior where droughts are a normal part of their environment. In addition to their ability to withstand stresses of heat and water shortage, the kangaroos cease to breed when conditions become extremely adverse. When conditions are favourable the build-up of the kangaroo population is aided by the ability of the females to carry a reserve embryo in the uterus at the same time that there is a joey in the pouch. The development of that second embryo is suspended while the first joey remains in the pouch, but it is born within a day of the first joey permanently vacating the pouch. So while the kangaroo can have only one young at a time, they can be spaced as closely as it is possible.

Macropus giganteus Great Grey, Forester
Recognition: Very large, head and body sixty inches, tail thirty-seven inches. Fur quite short, both sexes silvery-grey, snout hairy between nostrils. Face grey without light or dark markings.

Distribution: Eastern Australia from north-east Queensland (*M. g. giganteus*) to New South Wales and Victoria (*M. g. major*) extending inland as far as western New South Wales. Within this range the Great Grey is confined to denser scrubs and forests than the Red, whose distribution it overlaps by several hundred miles. The Tasmanian Great Grey or Forester is the sub-species *M. g. tasmaniensis*.

Notes: Like the preceding species, the Great Grey is a gregarious animal, and usually is seen in groups. It inhabits forested country, sheltering during the day among dense vegetation or boulders, and emerging at evening and early morning to feed in grassy clearings.

Red Kangaroo

Kangaroo Island Kangaroo

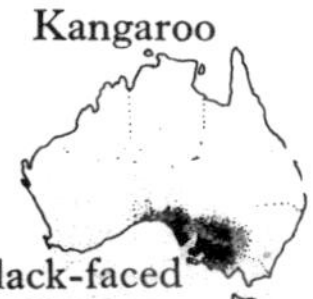

Black-faced Kangaroo

Western Grey Kangaroo

Great Grey Kangaroo

Eastern Wallaroo

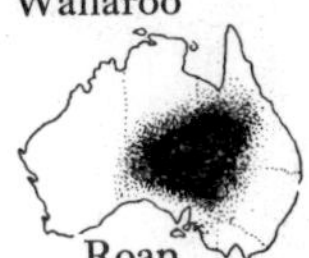

Roan Wallaroo

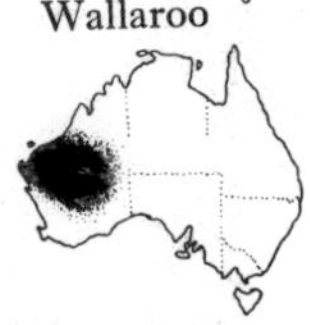

Western Euro

While the two mainland sub-species are very similar, the Tasmanian sub-species is of considerably heavier build, and somewhat more brownish colour.

Macropus fuliginosus other grey kangaroos
Recognition: Fur grey-brown to dark brown; face, particularly on the muzzle, darker brown. (Applies to all sub-species.)

Distribution: South-western Australia to western New South Wales (see notes on sub-species below).

Notes: Within this species are several sub-species which at various times have been classed as full species, and still retain individual common names.

Macropus fuliginosus fuliginosus Kangaroo Island Kangaroo
The first-named example of the species. More heavily built, very dark brown, lethargic and slow-moving. Inhabits Kangaroo Island, South Australia.

M. f. melanops Black-faced, Sooty or Mallee Kangaroo
South-western New South Wales, north-western Victoria, entire southern half of South Australia.

M. f. ocydromus Western Grey Kangaroo
Forested south-western Australia, extending eastwards through mallee country to link up with the preceding sub-species.

Macropus robustus Wallaroo, Euro or Hill Kangaroo
Recognition: Solidly built, with relatively shorter hind legs than the preceding kangaroos, which together with the thick long shaggy fur gives a stocky appearance. When hopping the body is held in a much more vertical posture than the Red or greys. Colour of various sub-species ranges from very dark grey to reddish-brown.

Distribution: The Wallaroo and Euro are found right across the continent, from eastern to western coasts, absent only from southern and extreme northern forested regions.

Notes: The distribution of this species is not continuous, but patchy, according to the suitability of habitat. In coastal eastern Australia, the almost-black Eastern Wallaroo is found on hill tops and rough parts of the Great Divide. Inland, where it is also known as the Euro, the species is found on stony ranges and breakaways, where the vegetation may be mulga or spinifex.

Studies of the Western Euro by Dr E. H. M. Ealey showed the species to be extremely well adapted to desert conditions. Like the Red Kangaroo, it can live in arid country wherever there is adequate water for its metabolic requirements. The Euro in particular can live on extremely poor-quality food such as the harsh spinifex. Euros have been observed to live as long as ninety days on a natural diet but without any water. They normally drink only in the hottest weather, and then as infrequently as once in fourteen days or even longer. They are probably equal to the camel in their ability to withstand dehydration.

There are numerous sub-species of Wallaroo and Euro, grading from dark grey in the east to rich red-brown in the west.

Macropus robustus robustus Eastern Wallaroo
Very dark grey; eastern coastal ranges.

M. r. erubescens Euro or Roan Wallaroo
Medium blue-grey with reddish tones. Central Australia, interior of Queensland, South Australia.

M. r. cervinus Western Euro
Rich reddish-brown; interior and north-west of Western Australia.

M. r. isabellinus Barrow Island Euro
Smaller, reddish. Confined to Barrow Island, north-west of Western Australia.

M. r. allegatoris Alligator River area, Northern Territory.

M. r. bernardus Northern Black Wallaroo
Small, dark. Rugged ranges of western Arnhem Land, Northern Territory.

Antilopine Kangaroo

Macropus antilopinus Antilopine Kangaroo
Recognition: Although this is one of the Euro group, its general appearance is rather more like the Red Kangaroo; the fur is less dense and shaggy looking compared with other Euros. It lacks the white and black muzzle markings of the Red. Colour is reddish-tan, paler below, with black on hands and feet; females sometimes greyish.

Distribution: Tropical northern Australia from north-eastern Queensland through the 'top end' of the Northern Territory to the Kimberleys.

Notes: Unlike the Euro and Wallaroo, the Antilopine generally avoids rocky-hilly areas, inhabiting tropical grasslands, open savannah-woodlands and northern eucalypt forests.

Wallabies

GENUS *Macropus*

Agile Wallaby

Macropus agilis (formerly *Wallabia agilis*) Agile Wallaby, Sandy Wallaby, River Wallaby, Jungle Wallaby
Recognition: Yellowish-brown fur, white hip and cheek stripes, short ears. One of the larger wallabies, head and body thirty-six inches, tail thirty-four inches.

Distribution: Tropical northern Australia from the Kimberleys to western Queensland.

Notes: The Agile Wallaby prefers the tall-grass country of the tropical plains, river flats and savannah-woodlands, where it feeds at night, retreating during the day into denser cover of scrub or gallery forests. The commonest of the far northern wallabies.

Macropus rufogriseus (formerly *Wallabia rufogriseus*). For common names refer to sub-species below
Recognition: A large wallaby, fur grey with reddish shoulders and rump, indistinct face stripe.

Distribution: Eastern Queensland, New South Wales, Victoria and south-eastern South Australia; Tasmania and Bass Strait islands.

Notes: The three sub-species are known by different common names.

Macropus rufogriseus rufogriseus Bennett's Wallaby
Smaller, stockier build than the other sub-species; longer and generally darker fur, reddish shoulders and rump indistinct. Found only on King Island, Bass Strait.

Red-necked Wallaby

Bennett's Wallaby,

M. r. banksiana Red-necked Wallaby, Eastern Brush-wallaby, Red Wallaby, Scrub Wallaby.
The mainland sub-species, with distinct reddish shoulders and rump.

M. r. frutica Brush Kangaroo, Bennett's Wallaby
Appearance as for the King Island form. Inhabits Tasmania and Flinder's Island.

Related Species:

Whip-tail Wallaby

Macropus parryi (formerly *Wallabia parryi*, *Wallabia elegans*) Whip-tail or Pretty-face Wallaby, Grey Flier, Blue Flier
Very long tail, white cheek stripe, white hip stripe going under tail from one side to the other.
Inhabits grassy sclerophyll woodlands, usually on hill slopes and summits, eastern Queensland and north-eastern New South Wales.

Macropus irma (formerly *Wallabia irma*) Black-gloved Wallaby, Western Brush-wallaby
A large wallaby with grey fur, black hands and feet, tail with black crest or brush on upper surface, yellowish cheek stripe. Inhabits the dry sclerophyll forests and scrubby sandplains of south-western Australia.

Macropus dorsalis (formerly *Wallabia dorsalis*) Black-striped Wallaby
A large grey wallaby with a distinct dark stripe down the centre of neck and back, and white hip stripes. Inhabits dry sclerophyll forests and woodlands from southern Queensland to inland north-eastern New South Wales.

Macropus greyi (formerly *Wallabia greyi*) Toolache Wallaby
Resembles *M. irma* but crested tail is pale-coloured. Generally regarded as extinct. Formerly occurred in south-eastern South Australia and adjoining part of Victoria.

Macropus eugenii (formerly *Wallabia eugenii*) Tammar, Dama Wallaby, Scrub Wallaby
A small wallaby, found usually in dense thickets of dry sclerophyll forests; there are four sub-species. The Tammar is found on the tip of Eyre Peninsula, on Kangaroo Island and other South Australian off-shore islands. In Western Australia it inhabits the south-west of the State, and various off-shore islands including the Abrolhos and the Recherche Archipelago.

Macropus parma (formerly *Wallabia parma*) Parma Wallaby, White-footed Wallaby
A very rare species inhabiting rainforests and scrubs of east-coastal New South Wales. Latest records are from near Gosford, previous record from Dorrigo.

Wallabies

GENUS *Wallabia*

Swamp Wallaby

Wallabia bicolor Swamp Wallaby, Black-tailed Wallaby
Large, stocky build, brownish-grey, belly rufous-orange and tail black. East-coastal Australia from eastern Queensland through eastern New South Wales and Victoria to south-eastern South Australia, in areas of dense moist forest, especially in gullies.

Pademelons

GENUS *Thylogale*

Small macropods whose hind foot is less than six inches in length, and tail comparatively short.

Thylogale billardierii Tasmanian Pademelon, Red-bellied Pademelon
Recognition: Long shaggy fur, dusky grey-brown but yellowish-orange on the undersurface, ears small, rounded and partly hidden in the fur, no distinctive face markings.

Distribution: Abundant in Tasmania; also occurs on Bass Strait islands and in southern Victoria, formerly in South Australia.

Notes: A gregarious species living in communities in gullies and damp places of low dense scrub and tall grass; its runways form tunnels through the tangled vegetation. Because of its abundance in Tasmania, there are control measures to protect farming and forestry interests; many of the skins are utilised by the commercial fur trade.

Related Species:

Thylogale thetis Red-necked Pademelon
Reddish shoulders and nape of neck. Rainforests and wet sclerophyll forests of coastal south-eastern and eastern New South Wales.

Thylogale stigmatica Red-legged Pademelon
Brownish fur, yellow hip strip, rufous heels. Indistinct stripe down the back of the neck. Head and body twenty-eight inches, tail fifteen inches. Eastern Queensland and New South Wales in rainforest and wet sclerophyll forest.

GENUS *Setonix*

This genus is distinguished from the preceding by the extremely short tail, very short ears and foot, and the unique dentition. The teeth differ from those of all *Thylogale* wallabies, and resemble the teeth of the tree kangaroos and the *Dorcopsis* wallabies of New Guinea.

Setonix brachyurus Quokka
Recognition: Small, head and body twenty-three inches, tail ten inches. Stocky build, tail extremely short and tapering, ears short and rounded, fur grey-brown.

Distribution: South-western Australia, in dense vegetation of swampy valleys in eucalypt forests. On off-shore islands including Rottnest Island.

Notes: The one species of this genus, the Quokka, has been of great scientific importance, being the first Australian marsupial to have its biology intensively studied. At the University of Western Australia, pioneering research by Professor H. Waring and Professor A. R. Main established the groundwork which has been the foundation of most later marsupial research. Discoveries included the processes by which the development of the embryo is delayed as long as there is a joey in the pouch, and the digestive specialities which make this and other macropods able to utilise the very low nutritional vegetation upon which they must feed.

Studies of the Quokkas in the field, at Rottnest Island, have revealed details of population fluctuations between harsh and favourable seasons, death rates, and requirements of habitat, food and water.

Nail-tailed Wallabies

GENUS *Onychogalea*

The three species of this genus have a small dark-coloured horny nail, like a fingernail, at the tip of their tails, where it is more or less hidden in the fur.

Their noses are strongly convex in profile, giving a 'Roman-nosed' appearance which is characteristic of all species. The tail is very long and slender, the incisor and premolar teeth are different to those of other wallabies, and all species have distinctive white cheek, shoulder and hip stripes. Differences between species are slight.

Onychogalea unguifera Northern Nail-tailed Wallaby, Karrabul
The shoulder stripe begins level with the armpit and extends down on to the chest. A common species across tropical northern Australia from north-eastern Queensland to the Kimberleys, found in savannah woodland.

Onychogalea lunata Crescent Nail-tailed Wallaby, Warrung
Rare. White shoulder stripe extends downwards from the base of the neck. Central Australia, north-western South Australia, and inland southern parts of Western Australia.

Onychogalea fraenata Bridled Nail-tailed Wallaby or Merrin
Rare. Shoulder stripe extends down and back from the base of the ear to the chest. Interior of New South Wales and Queensland.

Hare-wallabies

GENUS *Lagorchestes* and GENUS *Lagostrophus*

The name 'hare-wallaby' was given for their habit of nesting under tussocks of grass and low bushes and, when disturbed, dashing out with hare-like speed and leaps. Length, head and body about twenty inches, tail thirteen inches.

Lagorchestes conspicillatus Spectacled Hare-wallaby
A conspicuous orange ring around each eye. Common on Barrow Island, north-western Australia; occurs also in the southern Kimberley and Pilbara region of Western Australia, the Northern Territory, and the centre and north-coastal parts of Queensland, usually in spinifex.

Lagorchestes leporoides Brown Hare-wallaby, Eastern Hare-wallaby
Extremely rare, possibly extinct, the last specimen being taken in 1890. Black patch on elbow, rufous ring around each eye. Inland north-eastern New South Wales to north-western Victoria and adjacent South Australia.

Lagorchestes hirsutus Western Hare-wallaby, Warrup
Orange ring around each eye, long orange fur around hind-quarters. Bernier and Dorre Islands, in Shark Bay, and the interior of Western Australia extending towards Central Australia, in spinifex grasslands.

Yellow-footed Rock-wallaby

Northern Nail-tailed Wallaby

Brush-tailed Rock-wallaby

Spectacled Hare-wallaby

Lagorchestes asomatus
Known only from a single skull found near the Western Australia–Northern Territory border.

Lagostrophus fasciatus Banded Hare-wallaby, Munning
The thick fur is grey with transverse bands of black and white across the rump. Common on Bernier and Dorre Islands in Shark Bay, occurs also in parts of south-western Western Australia.

Rock-wallabies

GENUS *Petrogale* and GENUS *Paradorcas*

Rock-wallabies have become adapted for a life among the boulders, caves and cliffs of mountain ranges and rocky hills in many parts of Australia, both coastal and inland.

Their rock-dwelling habits have led to the development of certain features which are quite distinctive. The hind-feet are very well padded, with the soles of the feet roughly granular to prevent slipping on polished rock surfaces. The tails are long and slender, usually brush-tipped or tufted to serve as a rudder while the wallaby is airborne on long leaps across crevices and between boulders, and to act as a balancer on narrow ledges and precipitous slopes.

Petrogale xanthopus Yellow-footed Rock-wallaby, Ring-tailed Rock-wallaby

Recognition: A particularly beautiful species, with a distinctive colour pattern. The long untapering tail has a ringed pattern, the hind feet and forearm are yellow, and there is yellow and white on the large furry ears.

Distribution: Flinders and Gawler Ranges of South Australia, hills of far western New South Wales and south-western Queensland.

Notes: Specimens from south-western Queensland have rather indistinct tail rings.

Petrogale penicillata Brush-tailed Rock-wallaby

Recognition: Long, untapered tail dark and bushy towards the end. Many variations in colours and markings in the various isolated widespread populations. Head and body thirty inches, tail twenty-four inches.

Distribution: Throughout Australia in hilly and rocky localities; absent from Tasmania.

Notes: The Brush-tailed Rock-wallaby is exceptionally widely distributed, so that many of the isolated populations have become sufficiently different that they have, in the past, been described as separate species. It is now thought that many are not full species but geographical races of the Brush-tailed Rock-wallaby. There are now listed about twelve different sub-species.

It is probable that the Brush-tailed Rock-wallaby's exceptionally wide distribution, from cold south-coastal islands to tropical desert regions, has been made possible by the subdued and constant temperatures that prevail in the rock crevices and caves where they shelter during the day. In the caves it is up to 25°F cooler than the outside shade temperature in summer.

One of the most beautifully coloured and agile of all the sub-species is the Pearson Islands Rock-wallaby (*P. p. pearsoni*) found on a group of small rocky islands off the western coast of the Eyre Peninsula, South Australia.

Related Species:

Petrogale rothschildi Rothschild's Rock-wallaby
The fur is yellowish, and at certain times of the year the upper back becomes purple in colour. North-western Australia from Arnhem Land to the Kimberleys.

Petrogale godmani Godman's Rock-wallaby
Tail dark, becoming pale towards the tip. Mountainous eastern Cape York Peninsula.

Petrogale purpureicollis Purple-necked Wallaby
Purple upper back at certain times of the year. North-western Queensland.

Peradorcas concinna Little Rock-wallaby
Extremely small, head and body sixteen inches, tail thirteen inches. Back rufous-orange, tail thickly furred, untapering, darker towards the tip. Tropical northern Australia from Arnhem Land to the Kimberleys.

Tree-kangaroos

GENUS *Dendrolagus*

The two Australian species have come from New Guinea in the distant past at a time when that island was linked with Cape York. They have since then become sufficiently different to the modern New Guinea tree-kangaroos to be considered distinct species.

Both Australian tree-kangaroos are confined to the jungles of north-eastern Queensland.

Tree-kangaroos were once ordinary ground-dwelling kangaroos (and like all other kangaroos, descendants of primitive tree-dwelling marsupials) but returned again to the tree-dwelling way of life. They still have the basic kangaroo shape, but modified for tree climbing. The hind feet have become shorter and wider, with roughened padded soles for a non-slip grip. The sharp claws are stronger and more curved. The fore-limbs are much heavier and more powerful than those of ground-dwelling kangaroos, with large hands and claws.

The very long tails of the tree-kangaroos have never regained or developed any prehensile grip, but have a thick fur brush towards the tip serving as something of a rudder on long leaps, and as a balancer as they hop from branch to branch. The teeth have retained the browsing (leaf-eating) characteristics instead of the grazing modifications of other kangaroos.

Tree-kangaroos show amazing agility, and there are records of them jumping forty or fifty feet to the ground and landing, cat-like, on all fours, quite unharmed.

Lumholtz Tree Kangaroo, *Dendrolagus lumholtzi.*

Dendrolagus lumholtzi Boongary, Lumholtz Tree-kangaroo
Large, head and body twenty-six inches, tail twenty-six inches. Face, hands and feet dark, contrasting with grey back and very pale belly fur. Inhabits mountain rainforests of the Cairns and Atherton Tableland region.

Dendrolagus bennettianus Tcharibeena, Dusky Tree-kangaroo, Bennett's Tree-kangaroo
General colour brown with a dark rufous area at the base of the tail. Mountain rainforests between Daintree and Cooktown.

Musky Rat-kangaroo

GENUS *Hypsiprymnodon*

The one species of this genus is considered to be the most primitive member of the kangaroo family, still showing traces of their possum origin—the opposable toe, the scaly tail and the type of teeth.

Hypsiprymnodon moschatus Musky Rat-kangaroo
Recognition: A small bandicoot-like animal, head and body ten inches, tail seven inches. Front and hind limbs almost equal in length, and the first toe of the hind foot opposable to the other toes—the only member of the kangaroo family with the opposable toe. The tail is like that of the Scaly-tail Possum, naked of fur, and scaly.

Distribution: Rainforests of coastal north-east Queensland between Townsville and Mossman.

Notes: The Musky Rat-kangaroo runs quadrepedally instead of leaping like other kangaroos. It is a solitary animal, terrestrial and to a considerable extent diurnal. It is also unique among kangaroos in that it is partly insectivorous, obtaining insects, tubers and other vegetable material by scratching like a bandicoot among the debris of the floor of the rainforests.

Other Rat-kangaroos

GENERA *Bettongia*, *Caloprymnus*, *Aepyprymnus*, and *Potorous*

Western Brush-tailed Bettong

Bettongia penicillata Woilie, Brush-tailed Bettong, Brush-tailed Rat-kangaroo
Recognition: Small, head and body fourteen inches, tail twelve inches. Tail prehensile, crested with long black hair towards the tip.

Distribution: Formerly very widespread, from Western Australia's south-west and interior through Central and South Australia to western Victoria and western New South Wales.

Notes: Now extremely rare except in the forests of the south-western corner of Australia. This nocturnal herbivorous marsupial makes a grass nest in a hollow scratched out under a low bush or tussock of grass. The prehensile tail is curled around bundles of grass and sticks for transport of these to the nest site.

Related Species:
Bettongia gaimardi Eastern Bettong, Tasmanian Rat-kangaroo
Tip of crested tail almost always white. Formerly widespread, it has not been recorded on the mainland since 1910, but remains common in Tasmania.

Tree Kangaroos

Bettongia tropica Northern Rat-kangaroo
Crested tail may be either black-tipped or white tipped. This animal is distinguishable from the preceding species only by internal features. Eastern Queensland.

Boodie

Bettongia lesueur Boodie, Tungoo, Burrowing Rat-kangaroo, Lesueur's Rat-kangaroo
The tail of this species is not crested, but slightly thickened. Formerly very common from New South Wales, South Australia, Northern Territory, and much of Western Australia, but now exists only as a rare species on the mainland in Central Australia. Still quite common on Barrow Island, Bernier and Dorre Islands, north-western Australia.

Musky Rat-Kangaroo

GENUS *Caloprymnus*

Caloprymnus campestris Desert Rat-kangaroo, Plains Rat-kangaroo
A medium-small animal, head and body eighteen inches, tail

fourteen inches. Ears much longer than other rat-kangaroos, tail neither crested nor prehensile, front limbs very short. Inhabited sandridge flats and stony plains bordering the Simpson Desert, south-western Queensland and north-eastern South Australia, where it was last recorded in 1935.

GENUS *Aepyprymnus*

Aepyprymnus rufescens Rufous Rat-kangaroo

Medium-sized, head and body twenty-one inches, tail fifteen inches. Head and body rufous-grey, ears fairly long, tail tapering and neither crested nor prehensile. Coastal eastern Australia from Cairns to Newcastle, usually in eucalypt woodlands.

GENUS *Potorous*

Potorous apicalis Southern Potoroo

Recognition: Small, head and body sixteen inches, tail nine inches. Superficially resembles a bandicoot. Ears very short and rounded, tail short, tapering, prehensile. Hops with body held horizontally.

Distribution: Tasmania, islands of Bass Strait, eastern and southern Victoria, south-eastern South Australia.

Notes: The nocturnal Potoroos feed on roots and tubers which are dug up with the long claws of the forefeet. They do not at any time venture out of the dense vegetation in which they live.

Related Species:

Potorous tridactylus Potoroo, Long-nosed Rat-kangaroo, Gilbert's Rat-kangaroo

Resembles preceding species. Dense vegetation of coastal south-eastern Queensland, coastal New South Wales, north-eastern Victoria and formerly the south-western corner of Western Australia, (*P. t. gilberti*).

Potorous platyops Broad-faced Potoroo

A very small (rabbit-sized) species with a short head. Confined to the south-western corner of Western Australia where it is possibly extinct, not having been collected since 1875.

THE EGG-LAYING MONOTREMES, MONOTREMATA

SPINY ANTEATERS FAMILY TACHYGLOSSIDAE

The one and only Australian species which is covered with long spines, has a tubular beak-like snout, horny serrations on the tongue instead of teeth, reproduces by laying eggs which are carried in a pouch that forms during the breeding season.

GENUS *Tachyglossus*

Tachyglossus aculeatus Echidna, Spiny Anteater

Recognition: Spines, tubular bare beak-like snout.

Distribution: Throughout Australia and Tasmania.

Notes: There are slight differences between the Spiny Anteaters of different regions, especially the Tasmanian subspecies, which has shorter spines almost hidden in the longer fur. A very long-beaked genus, *Zaglossus*, occurs in New Guinea.

Spiny Anteater

FAMILY ORNITHORHYNCHIDAE

The only species of the family is extremely specialised for an aquatic way of life. The female Platypus does not have a pouch but in her nest burrow curls her body around the eggs to incubate them. The snout is superficially duck-like, but soft and leathery, combining the nose and lips of other mammals. The feet are webbed and the tail flattened transversely.

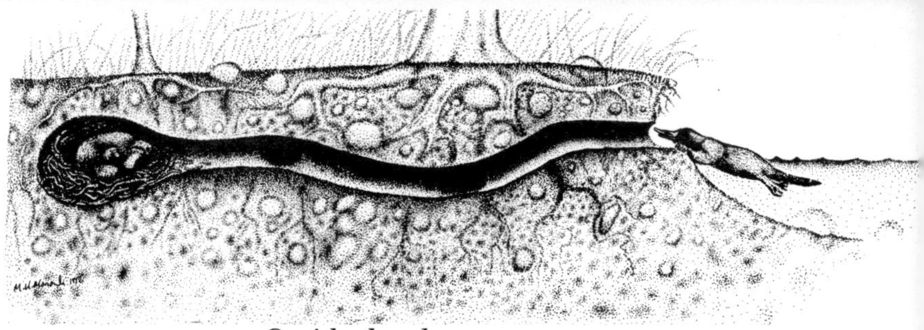

Platypus

Southern Potoroo

GENUS *Ornithorhynchus*

Ornithorhynchus anatinus Platypus

Recognition: Broad bill like that of a duck, dense brown fur, webbed feet, length, head and body eighteen inches, tail six inches.

Distribution: Freshwater streams and lakes of eastern Australia, from north-eastern Queensland through New South Wales and Victoria to south-eastern South Australia, also in Tasmania.

Notes: The Platypus excavates a burrow which is usually fifteen to thirty feet, but sometimes as much as ninety feet in length; the entrance is well above water level. At intervals the female plugs the tunnel with earth, probably to deter predators and to maintain a warm atmosphere for incubation.

A nest of grass and gum leaves is constructed, and two soft-shelled eggs laid. The Platypus, like the Echidna, has no nipples, the milk being sucked up by the young as it exudes from pore-like ducts of the mammary glands.

On the hind legs of both the male Platypus and male Echidna are sharp venom spurs. The venom of the Echidna does not seem dangerous to man but the Platypus venom has a very painful effect with swelling lasting several days.

RODENTS RODENTIA

AUSTRALIAN NATIVE RATS AND MICE FAMILY MURIDAE

This large family, wide-spread outside Australia, contains more than one-third of the rodents of the world. Members of the family have two prominent upper incisors which have enamel on the front surfaces only.

Before man introduced the most undesirable and often disease-carrying house mouse, ship and sewer rats of Europe, Australia contained almost fifty species of native rats and mice, all clean and quite attractive bush-dwellers. Some of these have been so long isolated on this island continent that they are considerably different to any rodents of other continents, while others have found their way to Australia comparatively recently, and closely resemble rats and mice of other lands.

Within Australia the native rats and mice have adapted to suit the great variety of habitats available. Naturally those earlier arrivals, known now as the 'Old Endemics', show the greatest degree of adaptation to Australian conditions.

Some have become suited to a life in desert regions. In our hopping-mice, *Notomys*, we have an excellent example of adaption to arid conditions as well as an interesting case of parallel evolution, for they have become remarkably like the desert-dwelling jerboas of Africa, although they have reached this kangaroo-like shape quite independently. At the other extreme, our indigenous water-rats have become very much specialised for an aquatic life in Australian streams. A great many species and genera of this family are endemic to Australia.

Many of the species of native rats and mice are difficult to identify, most genera and species being based upon features of skulls and teeth, requiring expert examination. However locality can be quite helpful in identification.

Australian Bush-rats GENUS *Rattus*

These are quite closely related to the ordinary introduced rats (i.e. the same genus) and in appearance and habits rather unspecialised. The various species tend to look much alike and are difficult to identify on external characteristics. It is in fact uncertain just how many species there are, and the knowledge of their biology is in most cases very slight. Recent studies of inter-breeding capabilities of supposed different species has recently led to the re-classification of some former species as sub-species.

Rattus fuscipes Southern Bush-rat

Recognition: Fluffy-looking brown fur, tail about the same length or slightly shorter than the head and body length.

Distribution: Coastal, from north-eastern Queensland through New South Wales and Victoria to South Australia and south-western Australia.

Notes: The Southern Bush-rat lives only in the bush, avoiding human habitation. It is a harmless shy nocturnal rodent that lives on native vegetation, and hides by day in a burrow. Its tunnel-like runways are often to be seen in low dense vegetation particularly around the borders of swamps and along creek banks.

Three former species are now considered to be no more than isolated populations of this one species; 'Southern Bush-rat' has been proposed as a single common name for all three. Those former species, now sub-species are:

Rattus fuscipes fuscipes Western Swamp-rat

Inhabits the south-west and some off-shore islands of Western Australia, usually in sclerophyll forest gullies and coastal heathland swamps.

R. f. assimilis Allied Rat

Coastal eastern Australia from north-eastern Queensland to Victoria, in forested and mountainous country.

R. f. greyi Grey's Rat

Coastal south-western Victoria, coastal South Australia west to the Eyre Peninsula, and Kangaroo Island.

Rattus tunneyi Paler Field-rat, Tunney's Rat

Recognition: Very short tail only slightly longer than the body alone; fur sandy-buff, whitish or yellowish below.

Distribution: Mid west-coastal and north-western Australia, through the 'top end' of the Northern Territory to eastern Queensland, parts of Central Australia and western New South Wales.

Notes: This very attractive-looking native rat inhabits a wide variety of country including savannah-woodlands, heaths, coastal sand flats and river plains.

Related Species:

Rattus lutreolus Eastern Swamp-rat

Relatively shorter tail, grey belly. Inhabits coastal eastern Australia from Queensland almost to Adelaide. Occurs also in Tasmania.

Rattus sordidus Dusky Field-rat, Northern Bush-rat

Northern and north-eastern coastal Australia from the Northern Territory to south-eastern Queensland.

Rattus villosissima Long-haired Rat

Grasslands of inland eastern Australia extending westwards as far as the Northern Territory.

Rattus leucopus Cape York Rat, Mottle-tailed Rat

Rainforests of north-eastern Queensland.

GENUS *Hydromys*

The water-rat is the only Australian mammal apart from the platypus to have become specialised for an aquatic way of life. These adaptations have occurred within Australia, making the genus unique to this continent. It has partially webbed feet, short dense fur, flattened head with nostrils well up and forward, high-set eyes, and very small ears.

Water-rat

Southern Bush-rat

Hydromys chrysogaster Water-rat

Freshwater rivers and creeks almost throughout Australia, Tasmania and many off-shore islands. Numerous races occur, many in the past being listed as distinct species—Western Water-rat, Atherton Water-rat, and others.

Related Genus:

Xeromys myoides False Swamp-rat

This and the preceding genus both belong to a sub-family different to all other rats. The feet of *Xeromys* are not webbed. Inhabits swamps of north-eastern Queensland and Northern Territory.

Tree-rats GENUS *Mesembriomys* and GENUS *Conilurus*

These large rats live in trees, nesting in hollow branches, and descend to the ground to feed.

Mesembriomys macrurus Golden-backed Tree-rat

Far northern and north-western Australia.

Mesembriomys gouldii Black-footed Tree-rat

Tropical northern Australia from Cape York to the north Kimberley.

Conilurus penicillatus Brush-tailed Tree-rat

The top end of the Northern Territory and the north Kimberley.

Conilurus albipes Rabbit-rat, White-footed Tree-rat

From southern Queensland through New South Wales to South Australia.

Tunney's Rat

Stick-nest Rats GENUS *Leporillus*

These large rats build massive nests of sticks on the ground, usually at the base of a bush or under a rock overhang.

Leporillus apicalis White-tipped Stick-nest Rat

Central Australia to western New South Wales and north-western Victoria.

Leporillus conditor Stick-nest Rat

From north-western South Australia to western New South Wales.

Rock-rats GENUS *Zyzomys*

Inhabitants of rocky localities; tails are long and thickened.

Zyzomys argurus Common Rock-rat

Northern Australia, from the Pilbara to north Queensland.

Zyzomys woodwardi Large Rock-rat

Far northern Australia from the Kimberley to Arnhem Land.

Zyzomys pedunculatus MacDonnell Ranges Rock-rat

Ranges of Central Australia.

GENUS *Mastacomys*

A relict species that was widespread in the distant past but now survives only in isolated alpine localities.

Mastacomys fuscus Broad-toothed Rat

Alpine areas of south-eastern New South Wales, eastern Victoria and Tasmania.

Mosaic-tailed Rats GENUS *Uromys* and GENUS *Melomys*

All species are distinguished by the mosaic-like pattern of scales on their tails, instead of the usual rings of scales.

Uromys caudimaculatus Giant White-tailed Rat

North-eastern Queensland, principally in rainforests.

Melomys cervinipes Fawn-footed Melomys

Coastal Queensland and north-eastern New South Wales, in forests.

Melomys littoralis Grassland Melomys
Grasslands of north-eastern Queensland.
Melomys lutillus Little Melomys
Cape York Peninsula.
Populations of *Melomys* occur in the Northern Territory, but it is yet uncertain whether these are separate, or belong to several of the above species.

Hopping-mice GENUS *Notomys*

In appearance these small mammals are like miniature kangaroos, but are not related to the kangaroos, being rodents, not marsupials. Nor are they related to the Jerboa Marsupial-mice (*Antechinomys*). The hopping-mice can easily be recognised as rodents by their chisel-like paired gnawing incisors, and absence of any pouch.

Hopping-mice have extremely long tails which have a brush tip that probably acts as a rudder to assist them to make sharp turns at speed—when they are airborne for a greater time than touching the ground. When pursued, they travel extremely fast for such small creatures, dodging between clumps of spinifex with very sudden changes in direction.

A great many species have been described in the past; however it now appears that there are not more than nine species and possibly less with further study and re-grouping. Most species are so similar in outward appearance that expert identification is necessary.

Notomys alexis Spinifex Hopping-mouse, Dargawarra

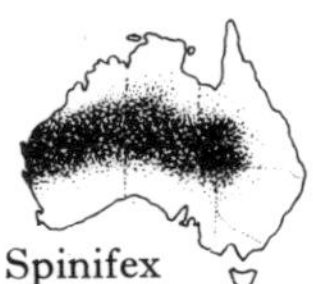
Spinifex Hopping-mouse

Recognition: About the size of a very large mouse, but with very long hindlimbs, very short forelimbs, extremely long tufted tail, and long ears. (Related species are similar.)

Distribution: Wide-spread through the interior of Australia, from western Queensland through Central Australia and northern South Australia to mid Western Australia. Prefers desert dunes and sandy spinifex grasslands.

Notes: The biology of this and several other hopping-mice has been the subject of intensive study. They are able to live in the arid deserts of the interior because they are nocturnal, avoiding the intense heat of day in burrows three or four feet deep. Although they will drink water if it is available, they are able to thrive, even gaining weight, on a diet of dry bird-seed without any water at all.

The burrows of the hopping-mice are complex and cunningly constructed. The author found their burrows in spinifex-covered sand dunes on the Kennedy Ranges, north-western Australia, and dug out several to obtain photographs.

Trails of tiny footprints in the sand lead to the main entrance hole, usually situated in the centre of a clear space between the spinifex clumps. But under, or close to, the nearby clumps of spinifex may be as many as nine escape holes, plugged with loose sand.

Upon excavating the tunnels it was found that the entrance burrow descended vertically, always reaching a depth of three or four feet, where there was a plug of soft soil, then the tunnel turned sharply to run horizontally. Several feet along this main horizontal tunnel was a nest chamber approximately a foot in length, four inches wide and three inches high. From the far side of the chamber were two escape tunnels, sloping upwards at first, then turning vertically towards the surface, where they were closed with sand.

In the centre of the nest chamber was found a cup-shaped grass nest, three inches in diameter, resembling the nest of a bird, but set into a hollow in the floor of the chamber. It contained three newborn hopping-mice, still blind and completely naked of fur. Their mother was found in one of the escape tunnels, digging her way up to the surface through the soft sand plug and packing the sand behind her as she went, as a barricade between herself and the intruders she could hear digging into her home. This tactic had been anticipated, and plugs of wood had been pushed into all escape holes that could be found. She was captured and re-united with her young, in due course becoming the subject of the series of colour photos of the young hopping-mice growing up in their subterranean home.

Related Species:
Notomys mitchellii Mitchell's Hopping-mouse
Woodlands, heathlands and grasslands of western New South Wales, north-western Victoria, South Australia, and inland south-western Australia.
Notomys longicaudatus Long-tailed Hopping-mouse
North-western New South Wales, central Australia and inland Western Australia.
Notomys aquilo Northern Hopping-mouse
Top end of the Northern Territory, and Cape York.
Notomys fuscus Dusky Hopping-mouse
From south-western Queensland to the south-east of Western Australia.
Notomys cervinus Fawn Hopping-mouse
Central Australia, and adjoining parts of Queensland and South Australia.
Notomys megalotis Big-eared Hopping-mouse
Very rare; Western Australia.
Notomys mordax
Extremely rare; Darling Downs of Queensland.
Notomys amplus Short-tailed Hopping-mouse
Northern Territory.

Native-mice GENUS *Pseudomys*

Resemble ordinary mice, but vary considerably in size. Shy, delicate, nocturnal, bush-dwelling. The native-mice, *Pseudomys* (meaning 'false mouse') are endemic to Australia and Tasmania, having evolved in this region over a long period of time. Their habits and distribution are for the most part not well known; most surviving populations are on off shore islands or remote or undamaged wilderness regions.

Pseudomys forresti Forrest's Mouse
South-western Australia, south-eastern Australia to southern Queensland, on heathlands and dunes.
Pseudomys delicatulus Little Native-mouse
Northern Australia from eastern Queensland to the Kimberley.
Pseudomys novaehollandiae New Holland Mouse
Eastern New South Wales.
Pseudomys fumeus Smokey Mouse
Forests of south-western Victoria.
Pseudomys australis Eastern Native-mouse
Inland eastern Australia, Queensland to South Australia.
Pseudomys fieldi
Central Australia
Pseudomys higginsi Tasmanian or Long-tailed Native-mouse
A larger species; rainforests of western Tasmania.
Pseudomys shortridgei Blunt-faced or Shortridge's Native-mouse
Western Victoria and south-western Australia.
Pseudomys desertor Brown Desert Mouse
Interior of Australia from north-western Australia to south-western New South Wales.
Pseudomys albocinereus Ashy-grey Mouse
South-western Australia, south-eastern Australia to southern Queensland, on heathlands and dunes.
Pseudomys hermannsbergensis Pebble-mound Mouse
Arid inland Australia from north-western Victoria to south-western Queensland, extending through Central Australia to north-western and south-eastern parts of Western Australia.
Pseudomys occidentalis Western Native-mouse
Heathlands of south-western Australia.
Pseudomys praeconis Shark Bay Mouse
Mid-western coast and islands of Western Australia.
Pseudomys gouldii Gould's Native-mouse
Southern Australia, from western New South Wales through South Australia to southern parts of Western Australia.
Pseudomys oralis Hastings River Mouse
North-eastern coastal New South Wales.

Pseudomys gracicaudatus Eastern Chestnut Native-mouse. Central and southern Queensland.
Pseudomys nanus Western Chestnut Native-mouse
Western Australia to the top end of the Northern Territory.

BATS AND FLYING FOXES *CHIROPTERA*

The only mammals capable of true flight, the bats—unlike the gliders—can propel themselves through the air with flapping flight comparable to that of birds. Their wings, like those of birds, are modified forelimbs, but the flight surfaces consist of membranes instead of feathers. These membranes are extensions of the body skin, supported and stretched tightly by the enormously lengthened finger bones that spread like the ribs of an umbrella. On some species these membranes extend to the hindlimbs and tail as well.

The considerable overseas research into bats has given an insight into many fascinating aspects of their biology, especially their radar-like high-frequency echo guidance systems. Their complex, and often very large, external ear structures are a vital part of the extraordinary equipment for 'flying blind' at night, funnelling in the faint echos bounced back from obstacles and airborne insects, giving distances and directions. Other elaborate folds of skin around the nose concentrate the ultrasonic bursts of sound they emit into a forward-directed beam. These intricate ears and noses are used as identification features.

Most bats are small and insectivorous, but there are also some very large fruit-eating species.

FAMILY MEGADERMATIDAE

A family of large bats occurring also in Asia and Africa. Australia has one species.

GENUS *Macroderma*
Macroderma gigas Ghost Bat, False-vampire Bat
Recognition: Very large, wing-span about two feet, length of head and body six inches. Fur ashy-grey, underparts creamy-white in desert areas, darker coloured in coastal part of range.
Distribution: Northern and interior regions of Western Australia, the Northern Territory and Queensland.
Notes: A powerful carnivorous bat which preys on small mammals, birds and reptiles. Budgerigars and Owlet-nightjars are the most common of birds taken. Ghost Bats hide during the days in caves and mine shafts, where they are very often overlooked because of their secretive habits and unobtrusive movements.

SIMPLE-NOSED BATS
FAMILY VESPERTILIONIDAE

Long-eared Bats GENUS *Nyctophilus*
The ears of these bats are joined together above the forehead; nose-leafs are not very large, often no more than rudimentary projections on the upper surface of the snout. Most species live in hollow trees, but some inhabit caves or hide in crevices under stones in open country. These bats are insectivorous, and will land on the ground to take beetles and other very small terrestrial prey.
Nyctophilus timoriensis Greater Long-eared Bat
Southern Australia, all States, and eastern Australia as far north as the Atherton Tableland.

Lesser Long-eared Bat, *Nyctophilus geoffroii*

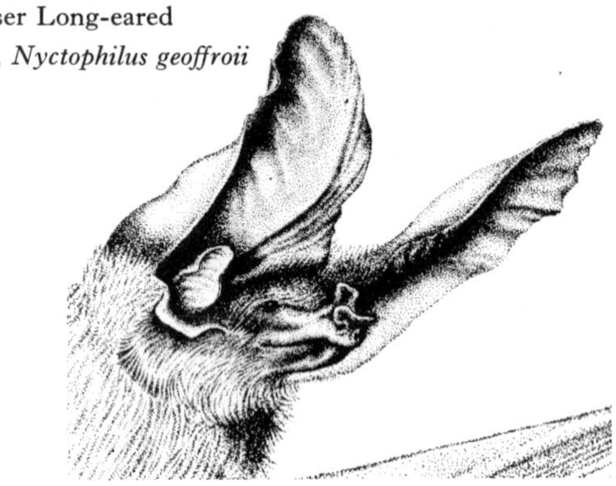

Lesser Long-eared Bat

Nyctophilus geoffroyi Lesser Long-eared Bat
Australia except the northernmost parts.
Nyctophilus bifax North Queensland Long-eared Bat
North Queensland and north-eastern parts of the Northern Territory.
Nyctophilus arnhemensis Arnhem Land Long-eared Bat
North-eastern Northern Territory.
Nyctophilus walkeri A specimen of unknown status, from the Adelaide River, Northern Territory.

Little Bat

Chocolate Bat

Ghost Bat

Small bats, various genera
No conspicuous recognition features, positive identification usually requires expert examination of the skull.
Miniopterus schreibersii Bent-wing Bat
Coastal northern, north-eastern, eastern and south-eastern Australia.
Miniopterus australis Little Bent-wing Bat
Eastern Coastal Queensland and north-eastern New South Wales.
Eptesicus pumilus Little Bat
Throughout Australia, living in caves, tree hollows, buildings, and the tunnel-like mud nests of Fairy Martins.
Chalinolobus gouldi Gould's Wattled Bat
Throughout Australia, hiding in dense foliage and tree hollows.
Chalinolobus morio Chocolate Bat
South-eastern and south-western Australia, and Tasmania, in tree hollows and caves.
Chalinolobus picatus Little Pied Bat
Western Queensland and north-western New South Wales, caves and tunnels.
Chalinolobus dwyeri Large-eared Pied Bat
Inland southern Queensland to central New South Wales.
Chalinolobus rogersi Hoary Bat, Frosted Bat
North Kimberley and the top end of the Northern Territory.
Pipistrellus tasmaniensis Tasmanian Pipistrelle
Coastal southern Australia, from south-eastern Queensland through New South Wales and Victoria to Tasmania; also in south-western Australia.
Pipistrellus javanicus Yellow-headed Pipistrelle
Widely distributed north of Australia; two specimens only recorded from unspecified parts of Australia.
Pipistrellus tenuis Timor Pipistrelle
One specimen only recorded from Australia, that from Dampier Land, west Kimberley.
Pipistrellus papuanus Papuan Pipistrelle
Cape York, Queensland.
Myotis adversus Large-footed Myotis
Northern, eastern and south-eastern Australia, from the Kimberley to South Australia.

Myotis australis Small-footed Myotis
New South Wales.
Nycticeius rueppellii Greater Broad-nosed Bat
South-eastern Queensland and eastern New South Wales.
Nycticeius greyi Little Broad-nosed Bat
All States except Tasmania.
Nycticeius influatus Hughenden Broad-nosed Bat
Central Queensland.
Phoniscus papuensis Dome-headed or Golden-tipped Bat
Eastern Queensland.

HORSESHOE BATS FAMILY RHINOLOPHIDAE and FAMILY HIPPOSIDERIDAE

These small bats have prominent horseshoe shaped skin structures around the top of the snout. These 'nose-leaf' projections probably play a part in their radar-like echo-location of objects in their flight paths. Possibly the nose-leaf channels the emitted ultra-sonic sound in a beam ahead of the flying bat.

The ears of the horseshoe bats are not joined together.

GENUS *Rhinolophus*
Rhinolophus megaphyllus Eastern Horseshoe Bat
Eastern Australia, principally coastal, from Victoria to Cape York.
Rhinolophus philippinensis Large-eared Horseshoe Bat
North-eastern Queensland.

GENUS *Hipposideros*
Hipposideros ater Dusky Horseshoe Bat
Far northern Australia, Cape York to the Kimberley.
Hipposideros galeritus Fawn Horseshoe Bat
Cape York Peninsula, Queensland.
Hipposideros diadema Large Horseshoe or Diadem Bat
North-eastern Queensland.
Hipposideros semoni Warty-nosed Horseshoe Bat
Northern Queensland.
Hipposideros stenotis Lesser Warty-nosed Horseshoe Bat
Far northern Australia, north-west Queensland to the Kimberleys.

GENUS *Rhinonicteris*
Rhinonicteris aurantius Orange Horseshoe Bat
Kimberleys and Northern Territory.

MASTIFF BATS FAMILY MOLOSSIDAE

Large bats, most with wrinkled squared muzzles like the Mastiff breed of dogs. All have a rather rodent-like tail which projects well behind the tail membrane. Some species commonly land on the ground, where they walk on hind feet and the clawed ends of the forearms.

GENUS *Tadarida*
Tadarida australis White-striped Bat
Southern Australia, from southern Queensland to southern Western Australia.
Tadarida jobensis Northern Mastiff Bat
Southern Australia, from northern Queensland to northern Western Australia.
Tadarida planiceps Little Flat Bat
Southern Australia, from New South Wales to Western Australia.
Tadarida loriae Little Northern Scurrying Bat
Northern Australia, from north Queensland through the Northern Territory to the Kimberley.
Tadarida norfolkensis Norfolk Island Scurrying Bat

Common Sheath-tailed Bat

Grey-headed Flying Fox

White-striped Bat

SHEATH-TAILED OR FREE-TAILED BATS FAMILY EMBALLONURIDAE

The tail, in this family of bats, has a surrounding sheath through the tail membrane, so that the tail penetrates and curves above the tail membrane. Their faces are sharp-nosed, rather like those of the flying foxes.

GENUS *Taphozous*
Taphozous georgianus Common Sheath-tailed Bat
Queensland, Northern Territory, mid and northern Western Australia.
Taphozous australis North-eastern Sheath-tailed Bat
North-eastern Queensland and Torres Strait Islands.
Taphozous nudicluniatus Naked-rumped Sheath-tailed Bat
North-eastern Queensland.
Taphozous mixtus Troughton's New Guinea Sheath-tailed Bat
Cape York Peninsula.
Taphozous flaviventris Yellow-bellied or White-bellied Sheath-tailed Bat
Throughout Australia, but not in Tasmania.

FLYING FOXES, FRUIT AND BLOSSOM BATS FAMILY PTEROPODIDAE

The four largest species, known as 'flying foxes', may have wing span up to four feet. Most bats of this family differ from the preceding small insectivorous bats in having plainly furred faces, without the complex folds of skin that serve to control emitted sounds, or the complex shaped ears. An exception is the genus of tube-nosed bats.

Most are fruit and blossom eaters, but some genera contain small partially insectivorous species.

Flying Foxes GENUS *Pteropus*
Apart from their large size, these animals can be recognised by their long-snouted, fox-like faces. They live in large 'camps', containing thousands of individuals, situated in rainforest, mangroves or swamp trees, flying out for distances of ten or even twenty miles in search of fruit and blossoms.

Pteropus poliocephalus Grey-headed Flying Fox
Recognition: Very large, grey, with a pale to reddish-yellow mantle around the shoulders and back of head, which is a lighter grey than the body.

Distribution: Coastal eastern Australia, in forests, from southern Queensland through New South Wales to Victoria, occasionally visiting Tasmania.

Notes: Although cultivated fruit is often taken, wild fruits of the rainforests, such a native figs, and nectar-filled blossoms of the eucalypt forests, make up a large part of their diet. Flying foxes are adept climbers, using the clawed thumbs of their winged wrists as well as their hindfeet. During the day they hang upside-down from the branches.

Usually only a single young is born, and carried each night to the feeding grounds until it becomes too heavy, when it is left at the camp and food brought to it.

Related Species:
Pteropus scapulatus Red Flying Fox
Reddish fur, indistinct orange-brown shoulder mantle. Coastal Australia, from eastern Victoria to north Queensland, the top end of the Northern Territory, the Kimberleys, and north-western coastal parts of Western Australia.
Pteropus alecto Black Flying Fox
Very dark brown or black, with dark reddish or yellowish mantle. Coastal Australia, especially in mangroves, from north-eastern New South Wales through eastern and northern Queensland, coastal Northern Territory and Kimberleys, to the north-west coast of Western Australia.
Pteropus conpsicillatus Spectacled Flying Fox
Black, yellow mantle, a ring of pale fur around each eye. Rainforests, north-east Queensland.

Spinal-winged Fruit Bats GENUS *Dobsonia*
Size similar to flying foxes, but distinguished by having the fur-less wing membranes extending right across the back, attached to the body along the spine instead of along the sides of the body. No fur on the back between the wings. Blackish colour, no mantle.
Dobsonia moluccense Spinal-winged Bat, Bare-backed Bat
Cape York Peninsula.

Tube-nosed Bats GENUS *Nyctomene*
Considerably smaller than the flying foxes, the tube-nosed bats have a wing span of about one foot. They are distinguished from all other large bats by the tubular extensions of their nostrils, but little is known of their way of life. There are many different tube-nosed bats on the tropical northern islands; two species occur in Australia.
Nyctimene robinsoni Queensland Tube-nosed Bat
Eastern Queensland.
Nyctimene albiventer Papuan Tube-nosed Bat
Eastern Australia, from Cape York to north-eastern New South Wales.

Blossom Bats GENUS *Syconycteris* and GENUS *Macroglossus*
Each of these genera has but one species, very small, and easily confused with the little insectivorous bats, from which they may be distinguished by the presence of claws not only on the thumbs (at the wing wrist joint) but on the index fingers as well.

Syconycteris australis Queensland Blossom Bat Eastern Queensland and New South Wales.
Macroglossus lagochilus Northern Blossom Bat Coastal Northern Territory and the Kimberleys.

THE NON-MARSUPIAL CARNIVORES
CARNIVORA

CANINES FAMILY CANIDAE
Australia has but one species of this family, and this is an animal so recently arrived from northern continents (probably accompanying Aborigines as a domesticated dog) that it barely qualifies as a native animal. It is in fact a well established introduced animal, a sub-species of the ordinary dog, *Canis familiaris*.

Dingo GENUS *Canis*

Canis familiaris (formerly *Canis dingo*, *Canis antarctica*)
Recognition: Not easily distinguished from many ordinary dogs of similar size except by a number of small features—the ears remain erect, tail is brushy, the canine teeth average somewhat larger, it utters yelps and howls instead of barking, and there are differences in basic behaviour patterns. Colour is variable, most commonly tawny yellow with a paler belly, white tail tip and feet.

Distribution: Throughout Australia, but absent from the more closely settled areas; absent from Tasmania.

Notes: The close relationship between the Dingo and ordinary domesticated dogs is shown by their ability to interbreed, but this probably occurs far less frequently in the wild, than commonly supposed. Variations in fur colour, often said to prove crossing with ordinary dogs, is a natural feature of the Dingo. Early explorers recorded seeing Dingos of a great variety of coat colours, in remote regions.

SEALS FAMILY OTARIIDAE
Many marine mammals inhabit the oceans around the Australian coastline—seals, whales, porpoises and the dugong, but only the seals emerge from the sea to spend any considerable time on the shores of Australia and its off shore islands.

Hair-seals GENUS *Neophoca*

Soon after birth the young of these seals loose their fine dense under-fur, retaining only a coarse hair coat.
Neophoca cinerea Australian Sea Lion
Recognition: Males very large, with pale yellowish manes. No under-fur on adults, dog-like faces, small ears. Hind flippers can be turned forward.

Distribution: From Kangaroo Island westwards along the south coast as far north as the Abrolhos Islands. Always on off shore islands except at Point Labatt, South Australia.

Notes: The male Sea Lion is one of the largest of Australian mammals, males exceeding ten feet in length, surpassed only by the Elephant Seal.

In spite of their size and awkward appearance, Sea Lions can move quite fast across a beach, and are surprisingly agile in climbing among coastal boulders. They can even scale steeply sloping coastal cliffs and have occasionally been found many miles inland. In the water they are fast and graceful.

Fur Seals GENUS *Arctocephalus*
Adult fur seals have dense fine fur beneath their outer hair coat. They have, in the past, been killed by the thousands for their fur. This industry was centred on the Bass Strait islands; sealing became unprofitable before all the seals were exterminated, and subsequently seal numbers have increased.

Compared with the much larger Sea Lion, the fur seals have a shorter and more pointed muzzle.

Arctocephalus doriferus Australian Fur Seal
Coastal Australia from New South Wales southwards to Victoria and Tasmania.
Arctocephalus forsteri New Zealand Fur Seal
The southern coasts of South Australia and Western Australia.

EARLESS OR TRUE SEALS FAMILY PHOCIDAE
More completely adapted for marine life than the previous family of seals, these have hindlimbs even more like the tail-fins of fish, making them more awkward on land. There is no external ear, and no underfur.

GENUS *Mirounga*
Mirouga leonina Elephant Seal
Even larger than the Sea Lion, males eighteen to twenty feet. Formerly a breeding colony on King Island, Bass Strait, now only a visitor to Tasmania and the south-eastern coast.

MAMMA

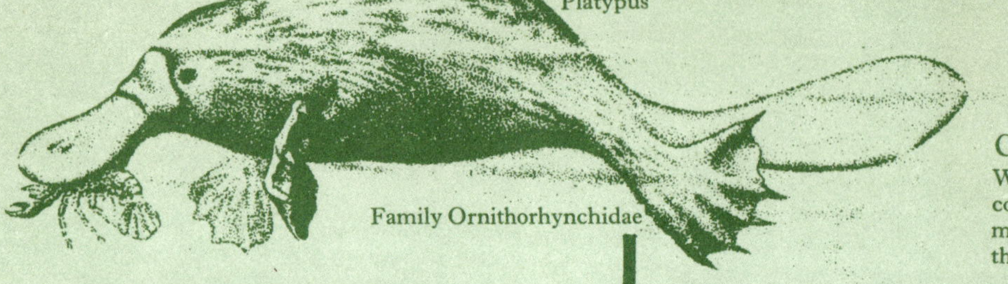

Ornithorhynchus anatinus
Platypus

Family Ornithorhynchidae

CLASS MAMMALIA

Warm-blooded animals whose skin is covered with fur or hair; females have mammary glands for the feeding of their young.

Family Tachyglossidae

Tachyglossus aculeatus Spiny Anteater

MONOTREMES

Sub-class Protheria

ORDER MONOTREMATA

Primitive mammals which lay eggs; there are only two surviving species.

PLACENTALS

Sub-class Eutheria

The young develop to a comparatively large size within the womb, nourished by the placental attachment.

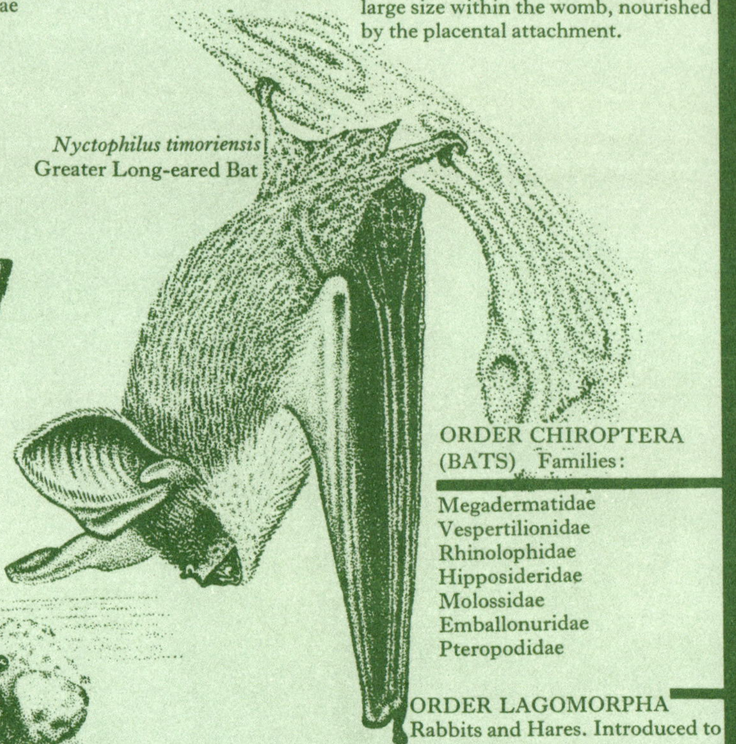

Nyctophilus timoriensis
Greater Long-eared Bat

Neophoca cinerea Australian Sea Lion

ORDER CARNIVORA
Families:
Canidae: the Dingo
Otaridae: Sea Lion, seals
Phocidae: Elephant Seal.

ORDER CHIROPTERA
(BATS) Families:

Megadermatidae
Vespertilionidae
Rhinolophidae
Hipposideridae
Molossidae
Emballonuridae
Pteropodidae

ORDER LAGOMORPHA
Rabbits and Hares. Introduced to Australia

ORDER PRIMATES
Man; monkeys and other animals not occurring in Australia

ORDER CETACEA
Whales: seas around Australi

ORDER SIRENIA
The Dugong; seas aroun northern Australia

ORDER ARTIODACTYLA
The even-toed ungulates: cattle, buffalo (introduced and wild in northern Australia), camel (introduced, wild in parts of central Australia)

ORDER RODENTIA
Family Muridae: native and introduced mice and rats
Pseudomys delicatulus Little Native-mou